About the Author:

Dr. Faye Tong obtained her MA in Journalism and Communication from Tsinghua University and Doctoral Degree of Philosophy from The Chinese University of Hong Kong. Moved into business from the academia, she writes the legendary story of business world with unique perspectives and styles.

# A LABEL, A LEGEND

FAYE TONG

The book was translated by Department of Applied Translation Studies in Beijing Normal University-Hong Kong Baptist University United International College.

SML
SML
www.sml.com

Coming by water,

riding over the tides of 1980;

Legend of label,

forged by 40 years of hardship.

# CONTENTS

# Foreword 1

After reading this book, I have a more rounded and deeper understanding of Simon Suen Siu Man, whom I have known for nearly 20 years. Four decades ago, he left the small village of Shangsha, Dongguan, in Mainland China for Hong Kong where he started another new challenging adventure. Simon has managed to overcome the ups and downs in his business and personal growth, fought persistently, and earned a wonderful life. Today, he runs a top-of-the-line label enterprise and generously supports cultural and educational causes. He sets up a museum, collects antique paintings and calligraphy, and has become a big fan of martial arts. He is a successful businessman who has lived a prominent life. Indeed, Simon is a man of wisdom, as commended by Prof. Jao Tsung-I. That impartial comment I personally cannot agree with more.

"A man of wisdom" is indeed a high regard. A clever man is not necessarily a wise one, just as the saying goes, "a clever person may become the victim of his own ingenuity". But a wise person must be clever. His wisdom is more about his mind, insight, and capabilities in aspects of life from socializing, decision-making to task-handling.

This biography focuses noticeably on Simon's personal growth and family life. His growth can be summarized as a "smile in tears, tears with light". He sees self-improvement as a life-long project and persists in working in a variety of ways to better himself. As a man who suits his action to words, he holds dear to the philosophies of truth-seeking, fact-pursuing, and value-persisting. He understands the strength and warmth of family and, as a result, values the family bonds. Simon regards managing family as a way in which he may pass down traditional cultural morality. He has proven that a harmonious family serves as the foundation for everything. A family is the basic unit of a society and is highly valued in China. The regulation of one's family represents the core of Chinese traditional culture, which views personhood as its base, self-cultivation as its core, and family order, national governance, and world peace as its basic thinking. Of course, the time has changed so much that we cannot just copy and paste the old experiences into modern society. However, the implied rationalities can still be constructive. He thinks highly of communal dining; for instance, he insists on having family dinners on all important days. It was believed by the ancient Chinese that etiquette originated from the dinner table. Dining itself is a form of communication and the repetitive practice of etiquette. As his children recalled, Simon attaches great importance to the "sense of ceremony" in dining. These principles have informed all aspects of his cultural concepts and ways of living. Simon roots his wisdom in the profound Chinese traditional culture. At the same time, he is a good learner whose quest for knowledge can not be satiated. He has benefited from Hong Kong which has opened up a whole new horizon for him, and his journey around the globe have made his world even bigger. Rooted in traditional culture and nourished by the essence of the human civilization, Simon has been constantly enriching his wisdom.

Simon's wisdom in business is fully represented in his philosophy that "business is busy with flexible ideas". The word, "busy", should not be overlooked as it is the essence, even the core, of business. Being busy means being dynamic, changeable and active. "Flexible ideas" is not just keeping up with the time but even forward-looking and forward-thinking. It is exactly what he means by saying that, "Emphasize the reality and recognize the changes, then change in response to change and change by virtue of that change."

It is easy to recognize with hindsight the significant moves mattering most in the development of a company. After these decisions, a company might be saved, improved, or expanded. What is remarkable about Simon is that he has different ideas and emphases informing each phase in the development of his firm, and he has made the right decisions to advance his company. When he just started the business, he "lost for now yet won the future"; when his company stood firmly in competition, he "channelled all their advantages to tackle one spot"; when his company grew, he "pioneered before others, excelled above others, changed ahead of others"; when his company became top in the industry, he aspired to "go global and become number one". Only someone with an ability to grasp the big picture can see far enough to seize opportunities, ease on temporary gains and losses, come up with unexpected thinking, sometimes even developing special and significant concepts, such as "reverse thinking". Thus, Simon makes seemingly impractical yet promising decisions. That is the power of wisdom.

The moment Prof. Jao Tsung-I said it was fate that brought them together, it was almost certain that Prof. Jao would exert a profound influence on Simon's life. Simon admires Prof. Jao as a world-renowned sinologist, while Prof. Jao praised Simon as an entrepreneur with a positive, amiable, and innovative characteristic. Simon showed his respect for Prof. Jao in promoting the Jao Studies as a sort of cultural consciousness, a homage to the Chinese culture. Jao had a notable impact on Simon, gently immersing the business leader in his erudition. Their relationship was a much-told story, going beyond age, occupations, and identities. Under the guidance of Professor Jao, Simon, already beginning a collection of calligraphy and paintings, became committed to the endeavour. He studied extensively the works of famous ancient Chinese calligraphers and painters such as Shi Tao, Bada, the "Four Wangs", "Four Monks", and "Eight Eccentrics of Yangzhou". As a result, Simon improved dramatically his knowledge, taste, and sensitivity in aesthetic matters. Simon has further grown his own taste and prefers collecting works from modern artists. The plaque in Simon's box room bears an inscription by Professor Jao, "Yi Tao Ju", is a testament to their relationship. The Sun Museum, represents a concentrated display of his artistic taste.

My personal connection with Simon dates back to 1999, when the China Cultural Relics Protection Foundation in Canada held a Chinese jade exhibition. The prime ministers from both nations examined enthusiastically the relics displayed in the hall where they met. In April of the same year, I was invited to visit Canada by the invitation of the Foundation, and the jades displayed were brought back to China by my colleague, Mr. Fan Shimin, who went with me. Later I discovered that Simon had sponsored this exhibition. Shortly after that, the Foundation also introduced me to Mr. Simon Suen Siu Man. By then, I was already working in the Palace Museum.

Simon was of the view that it should be a company's social responsibility and obligation to give back to society. He along with his company, benefited from such devotion. He funded the Chinese Museums Association (CMA) for the 2006 edition of *The Annals of China Museum*. The ceremony of his appointment as the Honorary Vice President of the CMA and the Honorary Chief Editor of *The Annals* and the launching ceremony of *The Annals* was held in October 2006 on the stage of the She Lent Fang Palace in the Palace Museum. As a representative of the organizer, I was fortunate enough to witness this important event in the history of Chinese museums. In a blink of eye, 14 splendid years have passed, and it is still vivid in my mind.

While congratulating on the publication of this biography, we look forward, and believe that Simon will make greater achievements and continue with his legend.

Professor Zheng Xinmiao
Former Deputy Minister of Culture
Former Director of the Palace Museum
Designated Research Fellow of China Central Institute for Culture and History
Director of Chinese Poetry Society

# Foreword 2

In this book, you are not just reading a book but also reading a man. A good book is one that touches people. And the main character of this book is indeed a moving figure.

Many Chinese from Mr. Simon Suen Siu Man's generation have suffered in their youth. Some managed to gain entrance into college after many years of farming, and, with the implementation of the Reform and Opening-up Policy, many more were able to fight their way up. Mr. Liang Qichao once said: "Adversity is the best way to learn". Likewise, Simon faced with members of his generation a similar set of adversity. Yet, it was his bravery and persistence that ultimately led Simon to become a successful entrepreneur and global business leader. There is no doubt that the nation-wide reform and opening-up policies were crucial. Policies aside, excellent entrepreneurs, such as Ren Zhengfei, Jack Ma and Simon Suen, cannot be neglected as part of the success story. These entrepreneurs started from scratch and strove constantly for success that contributed to the miracle of the Chinese economic take-off, saving the national economy and propelling the astonishing economic growth from the Cultural Revolution to an advanced status today.

What makes the story of Simon, a successful international entrepreneur, so touching is his profound humanistic care and his responsibility for the Chinese culture. I would like to cite two instances although ample examples have been provided in this book. At the beginning of this century, Prof. Jao Tsung-I decided to establish Jao Tsung-I Petite École at the University of Hong Kong. It is the norm in western countries and Hong Kong to establish a new research institution (or research unit) through self-funding. For example, social resources need to be raised for the operation of the Petite École. For that, Prof. Jao Tsung-I suggested, forming a Petite École Fan Club to support the Petite École financially. Simon, the biggest supporter and donator, is one of the founding presidents of the club. Not only has he made donations generously, but Simon has also encouraged a huge crowd of his entrepreneur friends to donate. It is fair to say that the Petite École would not have achieved so much without the assistance of Simon, who continues his support after the passing of Prof. Jao and held a grand fund-raising dinner in October 2019. The dinner turned out to be a great success, thanks to Simon's thoughtful arrangements and unrelenting efforts. The Petite École is now a place fulfilling the historical mission to revitalize Chinese culture — one of Prof. Jao's wishes.

Another example is Simon's full support for the establishment of the Jao Tsung-I Academy of Sinology at the Hong Kong Baptist University (HKBU) — an academy that has organized a capable research team, held a series of high-quality academic activities and published quite a few influential academic publications, thus advancing local study of sinology. Led by Prof. Chen Zhi, the Academy is also responsible for the Chinese-English translation of the key academic publications of Prof. Jao. I believe Professor Albert Chan Sun Chi, President Emeritus of HKBU, will provide a more detailed account in the foreword that follows this one.

In addition to his great talent and bold vision, global outlook, and humanism, I admire Simon deeply for his sincerity and modesty. Thus, it is no surprise that Simon has a wonderful family and many friends. The author of this book, Dr. Faye Tong, has been working with Simon for many years and knows all these details by heart, imbuing this biography with its vividness and appeal. The detailed description about the entrepreneur himself, and business management cases can be studied in high level business administration courses, forming an invaluable tool in the cultivation of business talents of the younger generation. I would like to express my gratitude to Dr. Faye Tong, for composing this inspiring and touching book with rich academic, literary, and historical value.

Professor Lee Chack Fan
Director of Jao Tsung-I Petite École, The University of Hong Kong
Member of the Chinese Academy of Engineering
Member of Guangdong Institute for Culture and History

# Foreword 3

I met Mr. Simon Suen Siu Man for the first time at the Scout Association of Hong Kong, when I was President of the Hong Kong Baptist University (HKBU). Dr. Wong Kwok Keung, the Honorary Doctor of HKBU, and some supportive friends introduced Simon who served as the deputy director of the Association.

Simon impressed me as a warm-hearted and humble person, who rarely talked about his international label business, and I was impressed even more when I learnt about the business scale. It was truly no small feat when someone can turn a business of a small label into a global enterprise.

Since we knew each other, it felt as if we were old friends. Although he was an entrepreneur and I had worked in a company before my university career, we never talked about business, instead focusing on shared interest such as Chinese culture and art. During our first lunch, he shared much about his collections and mentioned his favourite Chinese ancient painters and calligraphers: the "Four Wangs" and "Four Monks". I happened to have the painting collection of them. It turned out that Simon and I had a lot in common, and we enjoyed our conversations a great deal.

Being inspired and influenced by the world-renowned sinologist Prof. Jao Tsung-I, Simon has actively promoted the traditional Chinese culture and has made significant contributions to the growth of Jao Studies. He has provided invaluable support for the teaching of Chinese traditional culture at HKBU, establishing the Mr. Simon Suen and Mrs. Mary Suen Sino-Humanitas Institute Development Fund to further the operation, research projects and cultural exchanges of the Institute. With his generous assistance, we have been able to advance globally in Chinese Studies and Sinology. It was later that Simon participated in the establishment of the Jao Tsung-I Academy of Sinology, a milestone in the development history of HKBU. He is deeply esteemed as an entrepreneur who has taken profound responsibility for the social, cultural, and educational causes.

On the 28th anniversary of the establishment of SML, I wrote Simon, the founder, an acrostic couplet with his Chinese name: "少志立千秋，壯歲功成先報國；文心懷宇宙，華籤利就再雕龍". I praised him as a determined man who carefully carved out his way to success. The six characters, "少文功成利就" extracted from the couplet, standing alone, is my compliment and congratulations toward Simon and his company. And it is for the same reason that I am writing this foreword. Simon's success is closely attributed to his ambitions as a youth, his persistence as a young man, and his vision and courage — attributes demonstrated throughout his many experiences. All these characteristics are crucial for what he had achieved from a small label. Simon is admirable for a willingness to dive into an industry for three decades and reward the society as a successful entrepreneur.

I believe readers of this legend, from whatever fields, will be inspired to start and explore their own legends.

Professor Albert Chan Sun Chi
President Emeritus, Hong Kong Baptist University
Professor and Chairman of the Academic Commitee at Sun Yat-sen University
Member of the Chinese Academy of Science

❖ *Photo taken in Zhongshan Avenue, Shangsha Village in the 1950s. Suen Shing stood in the middle.*

CHAPTER

# I

# AMBITION

/

## BUILDING TRAITS

# Smile with Tears, Tears with Light

It was 1962, at the turn of spring and summer, Suen Shing, an accountant for a farm implements manufacturer in Shangsha Village, decided to return to Hong Kong, where he had spent his youth, to discover a new way of life.

Simon was sent to his grandaunt's foster care. He was less than five years old and was not yet aware of his father's leaving. Life has its own plans. Simon could not have imagined at that young age the impact of his father's trip on his future. No one could have dreamt that the son could repeat his father's journey, heading to Hong Kong where Simon would establish a label business from scratch. And, with the thoughtful application of one small chip, he would expand that business to the world and create a technology legend of a small label.

Simon began life in Shangsha Village in Chang'an Town in Dongguan City, Guangdong, a province in South China. It is the ancestral hometown of Mr. Sun Yat-sen, the first President of the Republic of China. According to the records of *The Genealogy of the Sun's Clan*, *The Pedigree of the Sun's Clan* and *Family's Ancestry Diagram of Prime Minister Sun*, at the Yuan Dynasty end and beginning of Ming Dynasty, the clansmen of Sun moved from Fujian Province to Shangsha Village, settling down and gradually growing into the largest clan in the village. Another village, Cuiheng, where Sun Yat-sen was born, was established as a branch of Shangsha. Sun Yat-sen was an eighteenth-generation descendant. At the entrance of Shangsha Community, next to the S358 Provincial Highway, there stands a memorial arch with an inscription written by Sun Yat-sen's grandnephew Sun Man — "The Hometown of Mr. Sun Yat-sen's Ancestors". The other side of the arch bears the inscription "Sun Yat-sen Avenue". Sun Man's younger brother, Sun Qian, inscribed these characters. The arch and Sun Yat-sen Avenue, which runs for 2 kilometres, were completed in 1948.

According to official accounts, Chang'an was established as a village in the Song Dynasty. At that time, this place had called Jingkang Salt Work due to its dependence on the industry. *The Jingkang Book* records, "People make a living by fishing, producing salt, and harvesting straw." Among the three, producing salt was a pillar industry. *The Food and Goods in History of Song Dynasty* records, "The thirteen salt works in the area generated revenue equal to more than 24,000 Dan (a measurement unit of weight and volume in ancient China) of grain per year. The profits from the trade were immense; the tax revenue alone was sufficient to pay for the soldiers' provisions throughout the South China."

The local salt industry began in the Song dynasty and continued until the Ming and Qing Dynasties. Chen Lian, a poet from Dongguan, crafted the following famed verse during the Ming Dynasty: "Cooking the sea with a thousand stoves burning, collecting salt with a thousand piles of snow stacking." The lines capture the scenes of prosperity during the period as well as the capacity of the Jingkang Salt Work to supply China with the life-sustaining condiment — a primary preservative of the day. The rise of the salt industry led to the development of the Pearl River Delta as an urban and trade centre, which also evidenced the Dongguan People's business acumen.

Chang'an, rich in diverse and unfolding terrains, is adjacent to mountains in the north and the Pearl River Estuary in the south. Shangsha, located in the heart of Chang'an, was once named Changsha and Shaxi, which should have been a bay in ancient times. After a period of illuviation, this bay became transformed into a fertile field. The evolution in topography may explain the village's name. Its inhabitants resided on a lengthy piece of sand by the sea. Shangsha means above the sand. An alternative explanation is that grains have a uniform size and texture that is uniquely suitable for farming. Shangsha can also in Chinese mean sand of premium quality.

Both definitions have merits. Shangsha is not only rich in natural resources but also gifted with fertile lands. It, thus, enjoys the reputation as "a land of fish and rice". The unique features of a local environment may endow its inhabitants with special characteristics. This is certainly true of Simon, who from his childhood exhibited an active and competitive character as well as innate cleverness.

Sun Jianping, Sun Juncai, and Sun Zhaorong, all veterans of the Vietnam War, were classmates with Simon in the same primary school in Shangsha Village. At a reunion of the class in 2017, Sun Zhaorong recalled Simon, as outstanding, displaying unique leadership skills from a young age that went unrivalled. However, Simon flashed a gift for playing tricks on his classmates. In short, he was a very naughty boy.

Sun Jianping, one year Simon's senior, lived in the village east and belonged to the same production brigade as Simon's. "At that time, on account of the Cultural Revolution, the schools were all closed. Although Simon was still young, he was very active. We set up children's corps, pretending we were fighting. I became the head. Simon took on the deputy. We made by hand the team flags, gathered many little friends and drilled on the grain-sunning ground of the production team. It's imposing."

Sun Jianping has a very good memory while describing vividly the pleasant past. It was as if the game had taken place only yesterday.

"At that time, Simon and I liked to play throwing stones the most, and would never be tired of it. We climbed up to the ridge of fields adjoining a 100-metre-wide rice paddy, playing with teenagers from the neighbouring Shatou Village, throwing stones at each other. It was considered a great victory if we hit our target. After a 'fierce' battle, when feeling hungry, we would sneak onto the farmer's reserved land to steal some sweet potatoes, then pick up some firewood and toast the treat. There's nothing more delicious than that taste."

While the head and deputy head bathed in a memorable smell of the sweet potatoes in their hometown, Sun Juncai, a classmate of Simon, suddenly asked: "Do you still remember Ms. Luo Junyu?"

"Of course!" Simon's eyes brightened up, recalling that afternoon a half a

❖ *At the class reunion held in 2017. From left: Sun Jianping, Simon Suen, Sun Juncai, and Sun Zhaorong.*

century ago.

At the time, the villagers were poor, busy earning a living that most didn't have time to prepare a lunch, frequently using the boiled water to reheat the previous night's leftovers. As usual, Simon casually swallowed a few bites and decided to go to school early. He bumped into Sun Juncai on his way. He blurted out an idea: "Let's go to find Ms. Luo."

As usual, Sun Juncai listened to Simon. The two teenagers walked excitedly to the teacher's dormitory.

Ms. Luo Junyu, the head teacher, taught Chinese. She was young and beautiful and was friendly to students. When they arrived at the dormitory building, Ms. Luo was sitting by the window, having her lunch. Simon grabbed a handful of clay from the ground and threw it to the window, trying to catch her attention. Unexpectedly, the clay fell right in her bowl. Ms. Luo poked her head out the window and checked, "Who's that?" Simon ran away immediately. Sun Juncai didn't run as fast as his clever friend, getting caught by Ms. Luo who rushed downstairs.

Some fifty years later, the recollection of the incident still inspires laughter amongst the small gathering of lifelong friends.

"Later, returning to school, I found Ms. Luo and honestly confessed my part. I also wrote a letter of apology. Yeah, a man must bear the consequences of his own acts!" Simon said.

One's personality is set from the early age. The straightforward and honest boy is unchanged.

Having calmed down from the excitement of reliving a childhood memory, Simon took off his glasses and rubbed his eyes with the back of his hand, wiping away the tears of joy. The naughty child, who ran fast that year, exhibiting a slight greying at the temples now, looked at his "accomplice" and said, "I am not sure whether Ms. Luo still remembers me." Sitting beside Simon, Sun Juncai put his wine glass on the table, his face a little flushed. He adjusted the sitting position, facing to Simon, saying, "She might forget anyone but not you"

A child born into a poor family gains the gift of maturity and responsibility. Such was the case with Simon, who was like any other child, would be naughty when he had the chance, but also needed to help with house and farm work.

Grandaunt relied on various means to feed her family, raising pigs and chickens as well as planting sweet potatoes and weaving hand protectors worn by farmer while they thrashed out the paddies to raise rice. He says self-mockingly, "I've known how to take care of my younger brother almost before I could stand." At the age of six, he was already helping his sister fetch water, cook meals, and feed pigs and chickens. "We could only raise one pig in a year. Selling the pig became the most important income my family would receive in a year."

Once for a while, he followed his grandaunt a few kilometres from Shangsha Village to the free market located in Chang'an where they would sell eggs, sweet potatoes, and hand-woven gloves. The first time going to the market with his grandaunt, he saw sachima: it had been cut neatly into small pieces and placed in a small glass-frame cabinet. He stared at the light-yellow colour of sweet sugar, his eyes firmly glued on the treat. He stood there and would not walk any further. Grandaunt pretended not to notice that and pulled him to go. He had always been sensitive to the appearance that he might be a guest under another's roof, but that time uncharacteristically, Simon made a scene. He was still a child and begged the grandaunt to buy the sachima glistening in the glass cabinet. Grandaunt sighed, before counting several cents from the coins and cash that she had carefully bundled into a handkerchief. Simon was full of joy, but his grandaunt had a mixed feeling, watching his hunger. Since then, no matter how poor they were, every time when they went to the market, the grandaunt squeezed out a few cents to buy sachima. "Actually, grandaunt loves me very much."

"My grandaunt's name is Zeng Di. She was the second wife of my granduncle. She married into Suen's family at a young age. Unfortunately, only a few years later, she became a widow. She raised alone my father and uncle, even though they were not her children." When Simon's father left their hometown in 1962, he entrusted Simon to her. "She brought up two generations of Suen's family. Her kindness to our family can never be repaid."

Simon became close to grandaunt. Although not well-educated, she would, every night, tell stories from *Water Margin* and *Romance of the Three Kingdoms*. Her storytelling was vivid: the chubby Lu Zhishen, who likes to drink and eat meat the most, is a pugnacious fighter. Zhuge Liang always waves a feather fan. It looks as if once he waves the fan, brilliant ideas will come out. Simon listened with relish and pestered grandaunt to tell one story after another. So that after he began to write letters to his father, the first thing he asked for was a set of China's four classic literary works.

At that time, Simon also lived with his father's first wife. Simon called her Da Ma, who relied on sewing clothes for a living. However, farmers made new clothes,

❖ *Photo taken in 1965. From right to left: Simon, his grandaunt, the younger brother of Simon, Da Ma, and the elder sister of Simon.*

at most, once a year. The income from the livelihood was very limited, and the work was not easy as well. The supply of cloth was tight, and depended on coupons. One piece of cloth may be needed to fashion clothes for the entire family, so Da Ma must carefully measure the cloth and make full use of the material. Even so, it was inevitable that the sleeves and trouser legs would be short. His trousers, Simon recalled, were invariably short. As a child, Simon knew but only vaguely that his father was in Hong Kong, yet had no idea about what he did exactly.

It was years later when he discovered that his biological mother had married his father when she was very young, but their marriage only lasted a brief time. The hardships that accompanied such an uneasy start gave Simon a drive and maturity that set him apart from his elementary education.

The school building of Shangsha Primary School, which housed the already mature young leader, was in the Ancestral Hall of Suen's Clan, a typical building of the Ming Dynasty. The building was fire-proof with reasonable ventilation. It had upturned eaves that were said to be shaped like the ears of an officer's cap, symbolizing a family's yearning for their offspring. Simon was at first an indifferent

❖ *Shangsha Primary School located in Ancestral Temple of Suen's Clan.*

student, his academic performance at best average. He didn't think he had much talent for studies. "Studying requires a calm mind, but I was hyperactive and naughty, so how could I sit still?"

Simon recalled that he was most afraid of mathematics. He could only score a 60 or 70 out of 100 in examinations, because he "couldn't calm down" when trying to work out the counting problems, so "always made mistakes." But Simon had a good memory and was capable of reciting. He didn't have a musical bone in body, so music class wasn't his favourite either. On top of that, he was often late to school, having to do house and farm chores. Once when he was late, teachers caned his palm three times with a ruler for punishment, but he disdained excuses. Sometimes he would skip classes, once even losing his school bag.

The teenager was not interested in becoming a top student, but he was determined to win each and every fight. The qualities of a leader inspired Simon to defend his friends and classmates, each admiring his heroism as well as his already strongly ingrained sense of justice.

Coupled with his intuitive capacity for management, Simon was able to lead his peers. He had ideas and plans, and it was only a matter of time before he became the most popular and influential classmate in the school. The group formed by Sun Jianping and Simon was so successful that what had been a small group soon grew large and was eventually under the school team. The teenager who lived in foster care was sensitive and competitive. The achievement brought a great sense of satisfaction to him.

His younger brother Suen Siu Wing, who grew up with him, recalled: "My big brother was so lively when he was young."

"Little boys from the different villages sometimes fought against each other. If our playmates from the same village were attacked, my brother would lead a group of friends to challenge the bully. Once, he was hit in the head by stones. Although injured and bleeding, he wasn't scared at all."

"In the autumn, not long after cutting the paddy when the weather was good and there was enough moonlight, my brother would gather the friends and ran together, in the wheat fields. The shadows of the teenagers were very long, as they jumped up and down in their bare feet. The wild night sky was quiet and clear, and the slow evening wind was filled with simple and real pleasure. We went very late into the deep night — tired but very happy."

These simple pleasures of childhood ended abruptly when Simon was twelve. A sudden storm completely changed his life.

On an ordinary morning in that summer of 1969, Da Ma asked Simon to go to Liaobu Town, 20 kilometres away, to take some steel bars back to Shangsha Village. As the "eldest" male in the family, he was obligated to perform heavy work from time to time. The weather was still clear when he set off, but suddenly it turned gloomy on his return trip. Simon was riding a bicycle alone with more than 100 kilograms of steel bars and foresaw that a heavy rain was coming. He couldn't help but speed up, hoping to get home as soon as possible.

The mountain path at that time was difficult to pass. Simon was only, by sitting on a hemp bag wrapped with rebar, able to reach the bicycle pedals which he drove relentlessly. But no one could have expected the torrential rains still came when he passed through Yangwu Village in Daling Mountain Town.

With thunder rumbling and lightning flashing, heavy rain pounded the boy's lithe body, individual droplets feeling like pricks of a needle. Alone, he was scared, but there was little choice. Simon pushed on courageously. The mountain path instantly became muddy in the heavy rain, and Simon slipped, the bicycle flipping over: its steel bars pressed against his chest, the sharp steel winger making a deep and sustained cut in his leg. The rain dripped into the wound. Simon lay in the tremendous onrush, feeling a stinging sensation which reverberated throughout his extremities. He was immobile and desperate.

Luckily, like the old saying goes, there's always a hope. Uncle Fangcheng, a villager from the nearby Yangwu Village, happened to be on this mountain path back home. Uncle Fangcheng brought him back to Yangwu Village where he allowed the boy to dry off. Then, once the rains ceased, he insisted on accompanying Simon back to Shangsha Village.

Meanwhile, the grandaunt and Da Ma were awaiting him, having become very distraught over Simon's long disappearance. It was dusk when they finally witnessed Uncle Fangcheng accompany Simon to the start of their village. The two of them thanked Uncle Fong Sing profusely for his kind help and were planning to boil ginger soup to warm up their absent child when the security officer of the brigade came to the house, asking to take Simon away. A reactionary slogan had appeared in a village alley, and a teacher had accused Simon of being its author. Simon was expelled from school and immediately imprisoned in a bullpen.

Simon felt in this moment as if his life was torn apart.

Only twelve years old, he was interrogated every day. Overnight, he went from being the deputy head of the children's corps, a position of great respect at his school, to a reactionary whom his classmates and teachers were free to criticize for whatever cause. Feeling aggrieved and wronged, Simon dared not ponder his future. Instead, his sole focus was on his present abject circumstance. Simon spent each night soaked in tears.

With his thin hands and feet tied with thick twine, he knelt on the cold black bricks of the courtyard of the ancestral hall, facing criticism from all teachers and students. A large placard hung from his neck, the cardboard wholly incompatible with his slight figure. The isolated and helpless teenager hung his head in fear and kept crying. The accusation sounded like sharp, rapid, and angry waves. Suddenly he felt a sharp pain before he blacked out...

In October, when his peers had entered the new academic year, Simon started his labour reform, carrying boulders, digging mud, and building dams on the side of the Dongjiang River. Living together with other peasant workers at a glass factory in a town, he started to work at 5 sharp each morning and left at 8 in the evening. When winter came, the weather turned cold. He carried a bucket and walked in bare feet along the river dam. He was frozen to the bones. Because of his small figure and poor sense of balance, he began to experience pain in his shoulders from carrying the buckets filled with stone. His grandaunt, nearly 70 years old, came to see him. Seeing the grey-haired aunt, he immediately fell to his knees, unable to restrain the flow of tears. Having encountered a boundless darkness prompted by unwarranted charges, Simon despaired and considered, ending his life. It was at that moment that he told himself that he had to keep going. While his peers were sitting in the classroom reciting "Before one's can achieve great, his mind must be toughened, his body exhausted, his stomach emptied, and his plan disturbed. Only in this way can he be motivated, strengthened, and developed." Simon experienced a profound recognition. Adversity strengthens the will while hardship tempers aspiration. The philosophy became Simon's personal template, his way of committing to strive after his dreams no matter the obstacle.

Two years after that fateful realization, he was returned to the first production team of the Shangsha Brigade where he became a farmer: seeding the rice, weeding, carrying the manure, and fertilizing the fields. He planted pumpkins, sweet potatoes, peanuts, bananas, and sugarcane, being diligent in completing each and every task well. In the morning, Simon arose and went to work on time. He felt a little dazed at first as he got up too early, but became "refreshed" by the cold water when stepping in the paddy.

Planting seedlings may look easy, but it was actually very difficult to perform the job quickly and efficiently over a long period. After bending for a long time, an unreal dizziness attacked when Simon raised his head. The thought of "Am I going to work like this forever?" would pop up. As the poem goes "Seedling a wide field, facing mud and dirt, my back is close to the clear sky." Though Simon had not read this sentence back then, he had already grasped an essential kernel: that in concentrating on one task, one achieves many. Soon, at an age of fourteen, he was able to plant seedlings as quickly as any adult man and was able, in one day, to fill more than acre with his plantings. "At that time, a primary labour earned a dozen work points a day, the equivalent at the time to six cents. But I could earn seven or even eight cents." Spcaking of the achievement, Simon exuded an understandable pride.

In 1971, Simon was assigned to a production team together with a number of young intellectuals. Feng Gaoyi, son of the well-known physicist Feng Bingquan and the opticist Gao Zhaolan, belonged to the same brigade. The educated youth took great care of their "little buddy". The labour camp allowed its residents to return home at managed intervals, and every time visiting Guangzhou, Feng Gaoyi would bring back some sweets and biscuits, sharing the rare and sumptuous treat with Simon. But what he felt the happiest about was the "food for thought" that the educated youths generously shared. From the exchanges, Simon grew to admire their knowledge and civilized character from the bottom of his heart. Besides Feng Gaoyi, there was also an educated youth named Chen Qinliu. This gentleman was very knowledgeable, having a high IQ and EQ: an unusual ability to support reason rationally and to deploy his logics to enhance his thoughts. Although forming a close friendship, the members of brigades often expressed sharp disagreement when talking over issues, frequently becoming red-faced. Chen Qinliu was different, maintaining a moderate tone when methodically analysing a subject matter and expressing his view with a clear and well-circumscribed logic. His calm display of intellect won the sincere admiration of everyone.

At that time, there was a popular saying: "Even if you don't learn the ABCs, you can still make a thing." However, because of his friendship with these educated youths who received education in English at the Affiliated High School of South China Normal University, Simon achieved a solid foundation in grammar. Chen Qinliu became one of his enlightening English teachers.

Simon started to practice of program by self-study, asking his older brothers to fill in the gaps in his knowledge. In the process, he began to tease out the nuances of grammar. Simon earnestly studied different sentence patterns and different tenses, although he didn't know at all whether one day his "present continuous tense" could turn into the "simple past tense".

In addition to learning English, he also became immersed in *The Selected Works of Mao Zedong*. Simon has an astonishing memory; a half a century later, he can still recall where, in each volume, an article can be found. From an early age, Simon became drawn to the majestic and boundless verse, like "Steel-liked mountains I fear not; climb over from its peak I stride." and "Seen all the world I stay young, for great views surround." Simon had a true hunger for knowledge, and his educated companions in the brigade would recommend books to broaden his intellectual horizon. He had achieved a bare literacy and would grasp only the central idea but as his vocabulary grew more expansive, he would reread the texts, grasping each time a new grain of knowledge. In the evening, he would become a rapt audience, listening in a small hut to these young intellectuals discuss the world. Simon found the exchanges inspiring, and soon, despite the heavy work of the day, his fatigue slipped off him. Overtime, he began vaguely at first to feel he must possess these ideals. The vision took root that life was more than a dream. It was a dream that could be realized. The ground was slowly being prepared, the winter planting nearly complete.

The spring finally came, and the country implemented new policies. After four years of labour reform, Simon was finally able to return to school and was assigned to the first grade of a junior middle school in Shangsha Village, Chang'an Town.

In the first maths class, he felt bewildered. The numerator and denominator, geometry, and equations were all Greek. After all, Simon had been out of school for four years, but he had been through too much and did not consider even the possibility of failure. Instead, he sought Sun Jingxian, a math teacher from Shangsha Village, for extra tutorials. Simon respectfully called this young teacher Uncle Jing. Although sympathizing with his situation, Sun Jingxian offered the blunt assessment: the gap was too significant, and Simon must return to the elementary school to consolidate his foundation.

But Simon believed that he would catch up, preparing a small notebook and would every night, study when others had gone to bed. Then, he would note down the places he didn't yet grasp, seeking help from teachers and classmates the next day.

There was no electricity in the countryside, so Simon would light a kerosene lamp. One night, he grew so tired that he accidentally fell asleep, almost starting a fire. Moved by the seriousness of his most diligent student, Sun Jingxian started giving out tutorials after school, discovering what Simon lacked in basic knowledge, he more than made up for with his powers of retention.

Simon learned the mnemonics of the Chicken-and-Rabbit problems very quickly: Chickens and rabbits in one cage; 35 heads they stretch; 49 legs the ground reach; how many of them each? Aware that using mnemonics best suited his student, Sun Jingxian was able to help Simon grasp rapidly the principles of mathematics. When teaching the concept of equation, the teacher says "Pork the mute seeks; don't know how much money there is; buy 50 grams then 0.1 dollar exceeds; if he want 100 grams then he needs another zero point six; how much pork can the mute take? The mnemonics worked very well. During his business trip to Tokyo in 2018, Simon was still able to recite these mnemonics clearly and fluently while chatting about the past with his local friend, Mr. Xu Yongzan, who had in the 1960s moved to Japan.

It was the same year when the US President Nixon officially visited China after the US table tennis delegation travelled to China, opening the door to Sino-US exchanges that had been halted for 22 years. That tour initiated what became known as the Ping-Pong Diplomacy.

After hearing the news in the countryside of Shangsha, Simon keenly realized that sports might also present an opportunity. He began to practice various track events: sprinting, high jumping and throwing. Turning to a habit first acquired as a farmer, Simon woke at 5 every morning to run distances. The halo cresting the moon had not wholly faded. The early morning was coated with quiet in the countryside when Simon appeared in a red vest and white shorts, circling a track dotted with dew. Many of his classmates retain an indelible impression of seeing Simon performing his laps. Because of his height, he recognized that his speed depended on frequency rather than the length of his stride. After school, he would use a similar pattern of strategic thought to develop a straddle technique, becoming determined after each fall to resume his efforts. Despite lacking in formal training, Simon mastered the essentials by carefully self-learning and self-reflecting every step and posture of his. Though failure was not rare at first, he improved his performance gradually. For a long time, he sprinted into his brand-new day and jumped out of it energetically.

Meanwhile, Simon was able to rekindle his relationship with his father in

Hong Kong, the two of them beginning a correspondence where Simon learned that his father had become engaged in trade, working more than ten hours a day. Still, Simon's father couldn't make ends meet. "I guessed he might be on the verge of bankruptcy. He was not in a good mood. I dared not ask more, but the idea of going to Hong Kong had already begun to take root." This was despite his father's frequent reminder that it was difficult to make a living in Hong Kong, and that survival was not as easy as Simon might think.

In the 1970s, Hong Kong's economy had not yet started to take off, and many of its residents still lived in wooden huts and tents. Simon fully understood that there would be no easy way for him even if he went to Hong Kong. Although only at the age of fifteen-year-old, he had already acquired a habit of taking a long-term perspective. He was well aware from his experiences as a child and as a farm labourer that the prospects would be limited in the countryside. "Staying here, I will probably be a farmer all my life. My idea is straightforward: as long as I have a job, I am satisfied. I have prepared mentally for the difficulties that my father described. I am self-reliant and self-confident, and I don't need to rely on my father. All I need is an opportunity."

Even Simon might not have expected the first opportunity to come so quickly.

On a winter day at the end of 1972, he overheard two fellow villagers saying that they were leaving that night. He immediately realized that they were going to Hong Kong. Without much thought, he decided to set off with them. The three teenagers rode bicycles for dozens of kilometres to the south end of Bao'an County. It got dark early in winter. Though not familiar with the surroundings, they somehow managed to get to Shekou and decided to swim the narrow distance to Hong Kong. They didn't know the river's name but later found out it was the Dasha River. They had not even reached the water when the border patrol caught them, cuffing and then sending them on the second day to the Shenzhen Detention Centre. Later, they were transferred to Dongguan Detention Centre. Simon lived on a diet of radishes in the Centre and would grow sick at the sight of that vegetable long afterward.

After half a year in detention, he was sent back to Chang'an. The moment he walked out of the Centre, the long-time-no-see sun shined so brightly that it hurt his eyes. His brother-in-law, Sun Heyi, came to pick him up, recalling that the first thing Simon did after coming back home was eat a whole bucket of rice.

The middle school principal in Shangsha Village wanted him expelled, but teachers all stood up for him. This was particularly true for Chen Xiang, the head teacher and Chinese teacher, and Sun Jingxian, the math teacher, who all believed that Simon is a hardworking and gifted student and should be given a second chance. As a result, Simon wrote a self-reflection paper and was allowed to continue school. It was at this point that his potential in sports became apparent.

Simon represented the school in the middle school athletics games in 1973, recording 12.4 seconds in the 100-metre dash, 1.57 metres in the high jump, and 65 metres in the throwing shot. These marks measured level three, a national standard for athletic accomplishment at that time. Simon's accomplishments caused an instant sensation. No one had ever thought that this 16-year-old middle school student would be so gifted in track and field events. What happened later proved Simon had made the right decision. As a student and talented athlete, Simon earned a ticket to high school. The Sports Group of Chang'an People's Commune even recommended him to a sports school after his high school graduation. But deep down, Simon knew that he wasn't meant to make athletics his career, so he refused it. He still wished to go to Hong Kong, where he saw broader opportunities for success.

"Because my father was in Hong Kong, and I had once attempted an escape, the authorities were certainly keeping an eye on me," Simon recalled. Militiamen would raid his home and search for his father's letters in the middle of the night. They tried to prove that he was attempting to flee again to Hong Kong.

A doctor from Heilongjiang had come to Shatou village, his son and Simon becoming close friends. The Committee in the village held a meeting to warn Simon: "They accused me of trying to gang up with people from other provinces and to resume my plan to flee to Hong Kong. But the truth is I didn't even know where Heilongjiang was." Many villagers shared a similarly negative opinion of Simon, believing the youth was misbehaving. However, they had misjudged the youth. "I had really wanted to be a soldier: then, to fight in the Vietnam War, just like Jianping, Juncai, and Zhaorong," Simon said. One family member serving in the army would make the whole family proud. However, his family background disqualified him as a soldier. Simon recognized that he had no other options: "I have no choice other than to escape to Hong Kong."

On the other hand, Simon wanted to meet his teachers' expectations, who fought for his chance to study. He was determined that he would surely complete high school. Young as he was, he had known that he should keep his commitment to people.

In the summer of 1976, Simon graduated from Chang'an High School, declining the offer of the People's Commune to enter a sports school and returning to the first production team in Shangsha, where he was given his first job to work at a duck farm.

Unlike a conventional farm job, a duck farmer was not exposed to sunlight or caught in a rainstorm when tending ducks. There was no busy season. Each day, he needed only to herd the ducks from a yard to the farmland and later ensure the ducks returned to the penned yard. After work, Simon lay down, at the grass, studying the sky. "So, is this how the rest of my life would be?" he mulled the matter. The sun, covered by clouds, would come out again, the light hurting eyes. He covered his face with a straw hat that was used to wear at work, dozed off.

"Simon, Simon!" Suddenly Simon heard someone calling him. He pulled away the straw hat and sat up to discover Sun Peiwen, the person in charge of education at the Committee of Shangsha village.

"Come back home! People from the Chang'an Education Office want to talk with you," Sun Peiwen said, running towards Simon, breathlessly.

Simon couldn't figure out why those people wanted him. Sun Peiwen, who stared at Simon, seemed to know but would not or could not disclose the answer. Simon kept quiet, following the leader back to the village.

When Simon arrived at his home, he met Li Xigui and Wan Yiguang from the Chang'an Education Office, "We want you to be a teacher at Chang'an Middle School and teach sports classes." They went straight to the point.

Simon didn't expect the news.

It was an absolute surprise to him. He had not majored in education or sports, and had never considered the possibility of becoming a teacher, not to mention a middle school teacher in the commune.

Teaching was and remains an honoured profession. When imagining the prospect of being called Mr. Suen, Simon grew excited over the opportunity. But this young man, who worked in a duck farm, also managed to remain calm at such a critical juncture and didn't accept the offer right away. Instead, he said, "I'll think about it."

Hearing the noncommittal response, Li Xigui said nothing. He stood up with Wan Yiguang. Before he left, Li patted him on shoulder and said, "Fine. Take some time to think about it."

When the Shangsha production team heard the decision, they were firmly opposed to granting the youth the opportunity. Some reported directly to Chang'an

❖ *Simon's high school graduation photo taken in 1976 (third from left in the second row)*

Commune that Simon remained a threat to flee to Hong Kong and was, therefore, not fit to be a teacher. The complaint was no small matter. Simon's new position required a change in production teams, and the Committee must approve all such changes. But the Education Office of the Chang'an Commune insisted that Simon should be transferred to the new post. Sun Quanbai, the person in charge of the production team in Shangsha, did not expect the Chang'an Education Office to be so determined. He finally relented, agreeing to permit Simon's transfer to Chang'an on one condition. The process had to be formal. In this way, Simon would be officially transferred to the Education Office of Chang'an Commune and would no longer be connected in any possible manner with the Shangsha village. Sun Quanbai remained concerned with Simon's previous attempt to flee to Hong Kong and did not want to bear any responsibility in case the future teacher should renew his plans.

Li Xigui and Wan Yiguang reported the condition to the Commitee of Chang'an Commune. Sun Xiqiu, the official in charge of education remained silent for a while before saying, "Okay, go ahead." Chang'an issued an official document, and Simon was formally transferred to a different production team. Simon never discovered why an official would take an interest in helping the youth. Sun Xiqiu used to participate in the interrogation when Simon was held in a cowshed as a counter-revolutionary in 1969 and might have felt compassion for the youth. For Simon, the job opened a window to a new life. Until this time, he had believed that he could only remain a farmer in the countryside and that his best shot was to go to Hong Kong and seek a job there. Then, the Education Office of Chang'an held a high opinion of Simon. And most of all, he felt pride in becoming a teacher, a respected member of the community. He decided to take the job.

In September 1977, a year after his graduation from Chang'an High School's agricultural and electrical class, having not yet attained his twentieth birthday, Simon had managed, through his drive and athletic capabilities, to gain a position, with the remarkable salary of 33 RMB each month. Villagers in Shangsha were all surprised, their feelings mixed with envy and admiration.

Two months later, the sports class was dismissed. Simon was reassigned to Chang'an No.1 Middle School to teach physical education and English. At that point, physical education was the major subject, while English served as the minor. Middle schools in the countryside had also begun promoting English.

Chang'an No.1 Middle School is co-organized by several production teams in Chang'an, including production teams in Shangsha, Xiaobian, Chongtou, Jinxia, and

❖ *Simon, was a middle school teacher, in 1977.*

Xianxi. Simon led a carefree life when teaching in Chang'an No.1 Middle School, describing it as a "merry poor life." Though having not much money, they were happy almost all the time. Often, Simon hung out with Chen Manlin, Li Bo, and several other teachers. Those talented teachers could play Erhu, a traditional Chinese musical instrument, and sang well. Every weekend, with some peanuts and rice wine, they would have a cosy afternoon. Those young people sometimes might try following the fashion together. For example, the most popular look at that time was to ruffle the hair until it looked like the folds of a chrysanthemom, wear a coloured shirt with large turn-down collars, and put on a pair of flared trousers. Chen Manlin recalled that the legs of the flared trousers were so vast, just like a horn. The legs were so long that the dust on the ground was swept away while they walked. They made fun of the trousers, laughing and saying that they were fashionable and helpful in cleaning the floor.

Simon taught very well, not only in physical education but also in English class. He taught English even better than some teachers who graduated as English majors. Grammar was essential in English learning, so what Simon had learned before had become a valuable tool in instructing his students who naturally, achieved good grades. He was very proud of that accomplishment.

It was in the same school that Simon met his wife, Mary.

Mary, a local from Xiaobian Village, also graduated from Chang'an Middle School. Having graduated with distinction, she was offered a teaching position in Chang'an No.1 Middle School. She taught Chinese and managed the general affairs in the school. Simon remembered how they first met. Mary suffered from a foot disease, so her colleagues went to visit her. Simon did not want to go at first, but Li Binghui, the School Principal, said it's not right to display a lack of sympathy to colleagues. "So, I went."

When he saw Mary's feet, which had become swollen and felt painful, Simon thought of a barefoot doctor he knew from the countryside, who specialized in treating such rare diseases. He suggested Mary should see Uncle Qi, advice which she took. Yet after seeing Uncle Qi for quite a while, and find Mary was not any better, Simon decided to look for a doctor in big cities. He took her to Guangzhou and asked his friends there for help.

Once Simon took up a task, he won't easily give up and did all he could to treat Mary's foot disease. "I took her by bike to many places in order to find the right doctor. As time went by, we spent a lot of time together, so love started to grow between us."

Simon was willing to help others, and he was a born leader. He displayed the same leadership qualities when spending time with other teachers. Since primary school, Simon had grown to be the leader. Similarly, in Chang'an No.1 Middle School, he was again the leader among other teachers. Young teachers liked to spend time with him. Unfortunately, the school principal disapproved of such bonding activities among young teachers and was firmly against Mary and Simon's close relationship. The principal believed such behaviour would undermine the authority of teachers. He made up some excuses and transferred Simon to another school, the Xinmin Primary and Secondary School in Xinmin Village.

Simon's colleagues in Chang'an No.1 Middle School missed him and still wanted to gather with him and, on weekends, would visit him at Xinmin village, a coastal village, some distance from Xiaobian village. The teachers had to ride a bike, and even had to cross one wooden bridge in the middle of the journey. The road in the countryside could become muddy and slippery in a rainstorm, which added extra difficulties. They need to carry the bike on shoulder when crossing the bridge. Even so, his colleagues went to visit him regardless of the weather. Mary was always with the crew.

In the bright moonlight, Simon felt warmth inside his heart. He was surrounded by love and friendship. However, sometimes while he was alone in this coastal village, surrounded only by the sound of the wind and the wave, that hopeless feeling for his future would resurface. Looking at the lights on the other side of the sea, Simon felt that he might still reach Hong Kong and seek opportunities there.

Quite a few colleagues shared a similar idea. Gao Mugen, a local in Xinmin village and a classmate of Simon's in middle school, was among them. The pair often discussed the possibility of "going to Hong Kong." Simon was determined to take Mary along, but Mary couldn't swim, so they would need to take the boat there. Fortunately, the Dongbao River ran through Xinmin village, a feature that allowed the plan to escape by boat a reasonable chance for success.

"There were altogether eight or nine of us. We had been planning for some time and had decided on the date of departure. Everything was ready. We just needed to wait for the right time to act," Simon recalled.

However, just the day before their action, on his way back to Xinmin Village after visiting his aunt in Shangsha Village, Simon was ambushed and detained by special agents from the Chang'an Commune. Later several militiamen escorted him to Chang'an Police Station, where he was charged with attempting to escape Hong Kong and with abetting others in such efforts.

He later discovered that Gao Mugen's mother had found out about their plot and reported the incident to Xigou, the Secretary of Xinmin village, who was fully aware of the severity of the incident and immediately reported the infraction to the Chang'an Police Station. Gao Mugen's mother had been warned not to speak to anyone else about the plan, in case Simon should try to escape.

During the trial, he remained loyal to his friends and refused to name the other participants in the escape. He only admitted that he had considered the idea but hadn't taken any concrete action.

"I've been grateful to my uncle, Sun Zhifang, all my life," Simon added. As a result of his work at the Wudianmei reservoir in Shangsha, Sun Zhifang knew a number of individuals who might be able to pull the strings. He set about trying to save Simon. At the same time, attempts to flee to Hong Kong were not uncommon. As the offense had grown less serious, people became used to it.

Still, Simon was detained in solitary confinement at the Dongbao checkpoint of Chongtou village for six months before finally, gaining his release.

"A return to Xinmin Primary and Secondary School was less than likely. They transferred me to Usha Primary School," Simon remembered.

From Chang'an No.1 Middle School to Xinmin Primary and Secondary School and then to Usha Primary School, Simon realized that he had no other choice but to go to Hong Kong.

He decided to make the third attempt, which nearly cost his life.

This time, he and his companions took a boat to Hong Kong. When they passed Fuyong, where the Shenzhen airport is now, the wind suddenly grew intense, and the waves were high. Simon knew they wouldn't make it to Hong Kong in such lousy weather, so they decided right away to abandon the boat and go onshore. The four of them decided to walk back home.

"Freeze! Raise your hand!" a group of border guards who came out of nowhere said, guns pointing at Simon and his companion.

It was late at night, so they were detained in the open space at the water plant near Shajing and were to be sent to the police station the second day. As he had been caught twice before, Simon understood that if he were to be sent to the police station and transferred to a detention centre, he wouldn't be set free again. He must find some way to escape.

He couldn't sleep all that night.

In the early morning, they were taken to Shajing police station in Bao'an County.

The police station was situated on a small hill about two or three floors in height. Nearby were some residential houses and narrow alleys. Simon studied his immediate surroundings while being escorted through the village. He walked into the police station and noticed the glass windows on the sidewall were open. The roofs from afar can be seen in the hazy morning sunlight. He had an idea.

The border guards uncuffed him and prepared to hand him over to the police. At that moment, he seized the chance to escape. He dashed immediately to the window and leapt out, landing on one foot while hitting the ground with the other, badly injuring his knee. He was bleeding badly. But Simon had no time to think about that. He stood up and ran as fast as he could.

Hearing the border guards chasing him, he had only one thought: Run! Run faster!

His knee was bleeding constantly. It seemed that the guards were closing in when he rushed into a house where a woman was sleeping. Woken by a stranger, she started crying out for help. Simon immediately knelt on the ground and begged her to save his life. As soon as the woman saw he was wounded and heard the guards outside, she knew what had happened. She quickly cleaned the blood from the ground with thatch and asked Simon to hide behind a pile of straw on the side of the house; she also gave Simon some straw paper to stop the wound bleeding. She must be one of those people working in the oyster farms in Shajing, so they had safflower oil in the house, a medicine often used by those farmers to treat wounds. When the guards could no longer be heard, she handed Simon a straw hat that the fishermen typically wore and told him to hurry off. Simon also dared not stay any longer for not wanting to bring more trouble to this kind woman. Before leaving, he knelt again and again to thank this woman.

He was unsure whether going by land was safe, so he swam across Dongbao River and returned to Shangsha village by the water. The seawater hurt his wound, and it was painful. It was a day and a night later when Simon was back at Shangsha village.

After several twists and turns in his journey, on January 17, 1980, Simon finally set foot in Hong Kong. He entered Hong Kong from Longgu Beach in Tsing Shan, Tuen Mun.

He never avoided the fact that he had been smuggled into Hong Kong.

"I tried to come to Hong Kong several times before. I have tried by water, land, and all other ways. I also experienced some dangerous, life-threatening moments. In 1980, I finally made it to Hong Kong by hiding in a boat," Simon said.

There was a galley below where many of us hid. Many couldn't stand the strong waves and felt sick and vomited. A terrifying thing happened when we were in the midst of the sea. The water suddenly started to leak into the boat. Many of us were frightened and tried to empty the boat of water. After a restless night, the boat arrived safely at Longgu Beach just after dawn. Many jumped into the water before the boat pulled into shore. But they didn't expect the water was still very deep. Some were not good at swimming. The currents simply took them away."

Simon recalled, "I've always been strong physically, but even for me, that journey was hard. By the time I got off the boat, my legs were shaking, and I couldn't walk. I asked my companions from the same village to leave first, but they wouldn't. They took me along. We hid in the mountain in Tuen Mun for three days and nights before going out, but we did not expect that would be caught right away by the snakeheads as soon as getting out of the mountain. They called my father and asked him to exchange me with 2,000 HKD for ransom. My father brought the money to rescue me. We then took a bus back to downtown. When we were passing Tsuen Wan, some policemen got on the bus to check. I was lucky because I sat in the front row, and they only glanced at the back rows. Noticing nothing strange, they just got off the bus."

Simon was lucky. On October 23 of the same year, a new policy was released, which would have prevented Simon from gaining legal status. By then Simon had just left his first job as a waiter, and took a new position at OCS, a Japanese courier service company. The young courier had never imagined that soon he would enter the label industry and that, due to his efforts and daring, would become a legend in the industry. Neither had he dreamt of standing in a university lecture hall one day and sharing his experience.

It was 5 PM on November 9, 2012. The audience packed in the Dr. Hari Harilela Lecture Theatre at Hong Kong Baptist University. Simon's public lecture, "Sharing on Business Management and Philosophy of Culture," began soon thereafter. He had been awarded the Honorary Doctorate in Social Sciences from the University in recognition of his outstanding achievements in industry and his significant contributions to public welfare, culture, education, and society. The honorary

❖ *In 2012, Simon gave a public lecture at Hong Kong Baptist University on "Sharing on Business Management and Philosophy of Culture."*

doctorate recipients would, as is customary, give a public lecture.

Simon, who had just celebrated his 57th birthday, stood on the podium. As usual, on the collar band of his black velvet suit was decorated with a red pin with his company logo. The pin might have been only 2 cm long; however, it symbolized the past three decades of continuous efforts. Spanning almost the entire back wall, a poster illuminated Simon in a beige suit on the poster, his hands crossed in front of his chest, smiling and calm.

Five hundred audience members were present that day. Beginning with his childhood experience, he then spoke of the years of labour reform, the ordeal that as a youth he had to overcome, the jobs taken after his arrival in Hong Kong, and his experience starting up a business. He talked about the inspiration, reflection, and encouragement that all these suffering experiences have since brought him. As a true optimist, Simon has never complained and never given up. He had always been grateful for being born in such a challenging time.

"What I've been through is a particular period when a lot of people have had unfortunate experiences and a lot of hardships. After so much suffering, a person needs great strength and courage to maintain his sincerity, honesty, motivation, and a positive attitude. What has been supporting me, along the way, is righteous thoughts."

He indeed possesses a noble ambition.

CHAPTER

# II

# ASPIRATION

/

## SETTING UP A BUSINESS

# Lose for Now yet Win the Future

Simon took three jobs during the period after his arrival in Hong Kong.

He began as a waiter at the Evergreen Restaurant, a Chinese restaurant in the Mira Hong Kong in Tsim Sha Tsui. His father found him this first job, paying monthly salary of 600 HKD. From studying advertisements on the street, Simon was able to find on his own a place to live. The place needed to be close to the restaurant. Being new to Hong Kong, if he had to navigate his way to work, he risked becoming lost. After scouring the streets hereabouts, Simon found the Hing Lung House near the Kimberley Hotel in Tsim Sha Tsui. He rented a unit with his little brother and cousin, who had come to Hong Kong with him. Not long after settling down, he found a part-time position at night as a dishwasher at a nearby restaurant. He had to work 15 or 16 hours per day. Even so, there was not much money left after paying the rent each month. But Simon was satisfied. He could feed himself and had a place to live. The life is plain but stable. In his view, it was more practical to guarantee survival first, and then, seek development. When he first came to Hong Kong, he had an accent but did not feel inferior. His life was poor, but self-reliance gave him a feeling of safety and security. After three months, he began to think about the prospects for development in the hotel industry and took the time to test and verify his judgment. He concluded five months later that a career in hospitality was not in his future. He quit and started looking for another job.

Simon's second job was as a courier at OCS, a Japanese courier service company. His idea was straightforward: if he wished to establish a career in Hong Kong, he had to know the streets. A job as courier provided an almost perfect vehicle to achieve the goal. He saw a vacant position in OCS, and applied right away, but the manager who interviewed him said no. The essential requirement was that the applicant should at least have finished high school education. Though Simon graduated from high school in Mainland China, he had no paper diploma, so the manager refused to hire him. If it had been another person, perhaps lacking in persistence, he would have given up. But Simon was not that individual. After sacrificing so much to come to Hong Kong, giving up was not an option for him. He also shared the manager's concerns and said honestly, "I didn't have a diploma at hand. But if you were kind enough to offer me a chance, I would accept lower pay during the one-month probation. If you are not satisfied with my performance at any time during the initial month, you can fire me right away."

What happened next might seem surprising, but not unexpected — Simon became a courier for OCS. He accepted a monthly wage 200 HKD below entry level, which was not a small sum at the time, but for Simon, that was the price of admission. "I got an opportunity with 200 HKD. It's a very good deal."

After the probation, Simon stayed, and the manager offered Simon a regular salary due to his excellent performance. Three months later, Simon was promoted and transferred from the Sung Wong Toi Road Branch to the head office in Central. "I remembered it was located on the 21st floor of the International Tower in Connaught Road Central. At that time, people thought you had to be really outstanding to work in Central," Simon recalled, smiling. "Even if you were a courier, you had to wear a tie and a shirt."

Simon, who oversaw courier service in Central, soon became familiar with the streets and the districts as well as with many businesses throughout Hong Kong. The opportunity, earned at a price, proved to be beneficial to his future career. Even though thirty years have passed since, Simon has retained a vivid sense of the period. Once he saw a delivery truck with an OCS logo in the parking lot, he walked over and asked his colleague to take a photo of him and the truck. It might be his way to pay tribute to his past.

While working at the restaurant, he had developed his skills in communicating with many types of individuals and working at OCS taught him more about the streets in Hong Kong. Once realized that he had reached the stage, and learning enough at the position, Simon knew it was time to move on. The ordeals that he had faced as a youth had made him very sensitive to changing circumstance. Simon also recognized the importance to reflect on and analyze a situation before determining the proper course. This approach allowed him to become successful, and gradually, he grew more and more confident, learning to exploit his strengths and avoid his weaknesses.

His work as a waiter and his experience dealing with front desk staff as a courier had shown his skill in communicating with all kinds of customers. That's why he felt he might be gifted in sales. His cousin, who also came to Hong Kong, knew that Fair Label was recruiting people and introduced him there. Though the pay was lower than a courier's, Simon thought that the position represented a good prospect. He decided to try his luck. He believed seizing an excellent opportunity for personal growth was way more important than earning in the short run a few more hundreds of dollars. He was not afraid of loss and more importantly, didn't view the decline in wages as a kind of loss. That's how Simon got his third job as a salesman at Fair Label.

In the 1980s, many bosses in Hong Kong were willing to hire mainlanders who had escaped to Hong Kong. These employers regarded the perilous journey as evidence of ambition. Hui Kai, the boss of Fair Label, saw that Simon was clever and hardworking and developed a high opinion of this employee.

Another worker at Fair Label, Lai Cheong, once mentioned that he would return home to visit family in the Mainland one day. Hearing this, Simon remembered his family back in Shangsha Village. The memory saddened him. He had gone through many obstacles to come to Hong Kong, thinking that he would never be able to return to his hometown. But no one ever knows what will happen in the future. He thought of going back to Shangsha to visit his grandaunt. It would be his first time back in his hometown since his arrival at Hong Kong. He couldn't just go back empty-handed. Simon asked Hui Kai for an advance on his salary.

Hui Kai agreed right away and wrote him a check of 3,000 HKD. But the check had to be cashed at a bank. Simon wanted to leave immediately and would not wait any longer. So, he gave Lai Cheong the check in exchange for 3,000 dollars in cash. Lai had been working at Fair Label for a long time and had some savings.

Lai had a similar experience and understood Simon's feelings. He gave Simon 3,000 dollars in cash. At that time, it was common for those returning home after a long absence to buy some household appliance. Simon bought a black-and-white television, which cost over 1,000 dollars. He also purchased some clothes and food. While carrying small bags with his hands and some big ones over his shoulders, Simon set off to Lo Wu.

"I still remember the sound of the train."

When he returned to Hong Kong from his hometown a few days later, Simon was still bathing in the joy of reuniting with his family. It was then that Lai told him the check could not be cashed.

At first, Simon thought that this was impossible. The boss had written the check. How could that be?

Simon realized afterward that the check was not backed with sufficient funds. Hui was unsure whether Simon would come back or not, and he didn't want to lose the money. Looking back, Simon saw it was an understandable precaution, but in his early twenties, he felt only insulted. Young as he was, he couldn't tolerate such shame. Rushing into his boss' office, slamming the check down on the desk, Simon said, "You didn't keep your promises. You do not deserve my respect. I quit!"

Hui Kai was good-tempered and never got angry. He graduated from Sun Yat-sen University in Guangzhou before arriving in Hong Kong. Sitting behind his desk all day, he counted every cent of his money and made every cent count; even kept his transportation fees in the book. Looking up at Simon, who was visibly angry, he said, "Simon, I was too busy lately, so didn't know that there wasn't enough money in my account. No hard feelings. You know me. I wouldn't do that to you on purpose."

"I hated those who did not keep their promises. And I was angry. So even though Hui said he would increase my salary and commission, I refused. I was determined to go," Simon recalled later.

He was furious at Hui at that moment. But he still respected Hui because Hui was the one who took him into the industry. He still called Hui his boss after his resignation. Hui passed away in 2018 when Simon was out of town. He made an overseas call to his wife in Hong Kong and asked her to represent him and bow in the mourning hall.

In those days, there were not yet regulations in Hong Kong against working for a competitor, but Simon felt that working for the competition was unethical. As a result, he had no work and no income for some time. When life got hard, Simon even needed to borrow money from friends. To save money, he used to stay in a "ripped room": a cell reaching to about 40 square feet, a bunk bed taking almost the entire area.

"I hung up a zippered bag to create a sense of privacy." His purpose in talking about the desperate conditions was not to complain. On the contrary, the stories were filled with his sense of humour. It is easy from these stories to recognize how these experiences shaped him into the optimistic and open-minded business leader that he would soon become.

Despite the severity of the conditions, Simon refused to bend his convictions, feeling it was not right to work for any other companies in the industry. Eventually, he decided the best course was to start his own business.

"Honestly, I never wanted to be a boss, but I didn't have a choice," Simon said. "I needed 1,000 HKD to open an account at the Bank of Communications in Yau Ma Tei. I had to borrow it from many friends to meet the sum."

Through a series of incidents, Simon started a small label concern, which would

❖ *Simon (left) visited his grandaunt (right) in his hometown and started his own business afterward.*

grow bigger and bigger, developing beyond his expectations into one of the world's leading brand label and solution providers.

Simon was only a broker at the very beginning. He accepted orders from those foreign firms and then, commissioned several factories to produce the woven labels. He invested his own time in exchange for profit. He recalled, "I just got myself a suitcase and then started looking for business." His experience as a courier proved useful. He knew precisely which foreign firms had a need for labels and where they were located. He started knocking on doors in search of opportunities.

"The first time I saw him was in the summer of 1981, in room 616 of the New World Centre, my company office in Tsim Sha Tsui," Raymond, Simon's good friend, remembered clearly how the two of them met. "He rang the doorbell and asked if we needed labels. It was hot, but he still wore a neatly pressed suit." Raymond added, "As we were a small firm then, our orders were usually for small amounts. There was a need to find someone to handle those orders. When he visited, we happened to have an order, so naturally, he got it."

However, that was a rare instant when Simon got an order so quickly and easily.

When Simon first arrived in Hong Kong, he spoke Cantonese with a strong Dongguan accent. While he spoke, people knew he was an "Ah-Chan" — a name for mainlanders in Hong Kong. Staff at those large foreign firms initially ignored him, but he was not discouraged at all. At lunchtime, he waited at the front desk; when they came out for lunch, he took the time to say a few words. If they didn't talk with him the first day, he would return the second day, then the third day. He showed up every day regardless of the weather. As an old saying goes, "Nothing is difficult for a willing mind." With such perseverance, he eventually gained more opportunities.

"At that time, it's not easy to get information about the Mainland. Those staffs in the foreign firms were curious. When I dined with them, I shared stories of my past and even about my embarrassments. Once, I shared a story about buying a watch from a Shanghai brand. The watch cost 120 RMB, and I had to save money for a long time. My monthly salary was only around 30 RMB. Then when I finally had enough money and bought the watch, it accidentally fell into the toilet. I had a hard time deciding whether to pick it up or not. They all laughed after hearing my story. A salesman has to act stupidly sometimes. If your customers like your story, they will speak with you again. That is one way you build up a relationship," Simon said. Later, it was not unusual for Simon to receive several orders even before he finished meals with those staff. His accent was supposed to be a disadvantage, but Simon somehow had managed to turn a supposed defect into an advantage.

After receiving orders, Simon had to find manufacturers. As most clothing and label manufacturers were in Cheung Sha Wan, San Po Kong, and Kwai Chung, he often went to "work" there.

Brokers need to cooperate reasonably with manufacturers to guarantee quality and timely delivery. Negotiations on the due dates of payment were most important above all. If brokers paid in a lump sum when purchasing the goods, they would never make money. So, Simon had to negotiate with the label factories to extend the due date of payment. In this way, he could first receive the goods, sell them, and then pay the manufacturers much later.

"It matters a lot, whether the due date of payment is 30 days or 60 days after receiving the goods. It might decide whether I could afford a proper meal that month," Simon commented.

Simon, who had never studied accounting, learned to manage finance based on his business practice. Even as his business expanded, he remained concerned about the turnover in accounts receivable and payable. Simon decided to focus first on a large factory in the industry, Man Hing Label Factory. He would each day follow Chan, the boss of the factory. Simon helped Chan run a lot of errands. Wherever Chan went, Simon would follow. Once, Simon followed Chan to Macau. As Chan spent a whole night playing cards in the casino, Simon just stood next to him all the time. He dared not close his eyes, nor leave. After quite a while, Chan knew Simon could be trusted and agreed to set the due day of payment 30 days after receiving the goods. Later, he changed it to 60 days.

In addition to Man Hing, Simon also made contact with several other big factories. He knew from the very beginning that it was risky to send all orders to one factory. However, if orders were given to several factories, Simon had to effectively distribute the orders. Therefore, he carefully gathered the actual production capacity of each factory in order to decide how those orders could be allocated. For example, the orders with a tight delivery time should not be placed on factories that already had "excessive orders." He knew how to find solutions to problems, and his solutions were convincingly reasonable. In this way, Simon gradually set up a network for meeting the requirements of his customers.

It is essential to have an expert in a specific position in modern enterprise management. The expert can supply guidance to ensure a detailed and rigorous division of labour. But in the 1980s, people in business who started from scratch had to shoulder many responsibilities on their own. One individual might be responsible for all: sales, delivery, and bill collection. Simon used to work like that; thereby, having to get up early in the morning. To this very day, he had kept to the same habit. His typical daily routine went like this: Simon went to Man Hing factory in Kwun Tong to pick up the goods; before he left, he'd remind the head of the factory not to forget the orders that would be due soon and to make sure of timely delivery; then he tidied up the packages of the labels and delivered those heavy goods to Marubeni in the Central Jardine House. After the delivery, he might also talk with the staff there to know more about the industry trends and to seek more business opportunities; at last, he went to Lark International in Ocean Terminal to collect payment.

"As a broker, you have to make sure the on-time reciept of payment from your clients and pay the manufacturers on time as well. Otherwise, there will be a huge problem. So, you have to choose your clients very carefully," Simon pointed out.

Despite Simon's caution, things didn't always go as smoothly as planned. Once, when he didn't receive a payment on time, he sat at the door of that foreign firm for ten hours. Then he felt a stinging pain. His palms had grown swollen and purple from holding onto a rope that had been used for packing the labels. There were many cases like this, but Simon only mentioned these incidents occasionally.

In 2007, when his business had already become well-known, the "Hong Kong People" program of CCTV 4, an international channel of China's national central television, did a special interview with him. He shared one of his experiences: Once, he delivered an order, but unexpectedly, the clothing factory which received the goods, closed soon after the delivery. He couldn't get the 50,000 HKD owed to him. It was a considerable amount of money back then. He was so anxious that he couldn't sleep at night. During the interview, he looked serious, spoke clearly, and pronounced each word accurately. He said, "I was very frank with the label manufacturing factory. I tried to have the boss understand the situation and agree for me to pay back the money in installment." After he said he finally paid the money back, the camera turned to a shot of him and his team inspecting a workshop that manufactured products on an extensive scale. He looked relaxed on the television image that accompanied Simon's story. It is common for businesspeople to go through many obstacles and experience sleepless nights under a guise of an apparently ordered daily routine.

"I started as a broker. An agent had an office while a broker didn't," Simon explained. He made this very clear all the time. It was not until later that Simon rented a place in Tai Kok Tsui to officially start his label company. His company was named "Hap Seng", meaning "success can be achieved with a concerted effort".

At first, his company could only afford one office. Fellow villagers who came to Hong Kong from Chang'an helped, painting the walls and waxing the floors in order to save money. Simon borrowed 9,000 HKD for decoration, a sum he paid back later. The office desk had rusted corners. It had been refinished from furniture left on the stress. Simon said, "The desk was still usable. We just needed to clean it up. That saved us some money."

Thrift was a virtue that came natural to Simon. The telephone was the only modern equipment in the office. That's it. Simon and Mary were married by then. The two of them worked together in the office. Simon remembered, "My wife answered the phones. I oversaw sales and bill collection. We hired a guy called 'Big Suen'. He was responsible for delivery."

Both the foreign firms and the manufacturers were his clients, which is another way of saying they were his bosses. "Those foreign firms do not care who handles their orders, as long as the goods were delivered in a timely manner and were of a

❖ *Simon and Mary started their business as husband and wife.*

high quality. Each broker and agent might also approach the same manufacturers to produce the labels. A key to business success was to establish a high level of cooperation with foreign firms and manufacturers and build up your reputation." Simon knew how to put himself in another person's shoes.

As Raymond, a business partner and friend for many years, remembered that Simon worked hard and put the clients in the first place. The attitude greatly impressed him. He said, "Simon delivered quality production on time. I started to introduce more clients to him. He had hired someone to help but would sometimes deliver the goods himself. Simon is a sincere and responsible person."

"I accepted all the orders that came in because above all, having enough food was what mattered most," Simon explained.

To Simon, sales volume was the food for any company. In addition to woven label, he accepted other kinds of accessory orders, such as leather badge for jeans, cartons for packaging T-shirts, plastic boxes, canvas bags, etc. Once recieving an order, he started searching for manufacturers. Factories of plastic products were mainly located in Tuen Mun and Yuen Long. However, in the early 1980s, it took longer to travel to Yuen Long and Tuen Mun from downtown than it did to reach Guangzhou from Hong Kong now. The round trip might take a few hours. However, he believed that the orders from a new industry represented a new opportunity to expand his business. If he could provide one more type of service to his clients, there would be one more avenue for the future growth of the company. Many experts in the industry sneered at his "cross-boundary" approach — taking an order regardless of its relevance to his primary service. Still, due to this unusual method, his company acquired a reputation for a can-do attitude. Customers began to approach him, and the number of orders he received rose considerably.

As the order volume increased, some customers wanted to visit the factories where Simon had their goods manufactured. Simon did not own a factory. So, he came up with the idea of setting up "joint factories." The scheme amounted to a supplier allowing a factory to be a customer showcase. In return, Simon would make sure to steer stable orders the supplier's way. It was a win-win for both parties. Simon explained: "you have to rely on experts to develop."

The model was very successful. At its peak, Simon's first company had established ten joint factories with different suppliers, including some of the most significant in the label industry.

Thus far, we have seen examples of Simon's daring as a businessman, but he was, by no mean, a fool. He knew how to analyze a situation and to determine the best opportunities for future growth. Simon recognized that if his business was to reach its full potential, he would need to produce labels independently. As long as he relied on the suppliers for production, the amount of profit would be limited, but if he owned the plant, he could pocket the entire sum. Of course, running a factory had the potential to drive up the operating costs, so there was some risk involved, but after carefully considering the situation, Simon entered into his first venture, acquiring a small printing factory, Zhi Fu.

In the 1980s when Hong Kong's economy was just beginning to take off, the garment industry and its upstream and downstream sectors were in full swing. While the economy was growing steadily, Simon was making ambitious moves. In 1982, he concluded a large deal with a quality client, earning thirty thousand HK dollars.

"I received a quotation request for a large order of domestic sales. Normally a quote is estimated based on the specifications given as part of the order. Thinking that since the chance of winning ranged from slim to none, I simply disregarded the normal estimate and was bold enough to ask for a high price to aim at big returns.

To my surprise, I won the order. Then, once the order was delivered, I had to wait anxiously. My fear was that I would not receive payment. So, I went to the office building where the buyer was located everyday, to check on whether the staff, as usual, returned to work. I would speak with the receptionist whenever had the opportunity, just to ensure the company was still in operation."

"When holding the cheque, my hands were shaking." Talking about his very first "big" deal, Simon said it was unrealistic to expect such high profits for every deal, so one should learn to be steady, to win the race. But when the opportunity comes, one should be bold enough to win.

By being "bold enough to win", Simon made his first pot of gold.

He used the money to buy a unit at Cheung Sha Wan. This site became the office for Zhi Fu which handled the screen-printing business. The joint factories, on the other hand, mainly handled the woven label business. The Hap Seng Company in Tai Kok Tsui continued to engage mainly in trade. Despite the small scale of his business, he made a conscious effort to delegate, when possible, the responsibilities and powers to his staff: thus, managing the operation in an orderly manner.

"When I came to Hong Kong, I was a penniless village boy. Every time I stood at the foot of the Lion Rock, looking at the Victoria Harbour and the starlight on the Victoria Peak, I had a strong desire to make a career here. There is a saying: to fight to the bitter end. That's the athletic spirit. I've been there. I've experienced it so I understand it very well." He described himself as a man who "knows how to fly if given the chance".

If Simon's move from employment to self-employment was a choice made out of no choice, his transition from trade to manufacturing represented new beginning and, as well, an opportunity for Simon to take stock of the situation.

In 1983, the Chinese and British governments began formal negotiations on the handover of Hong Kong to China. This was in a five-year period after the reform and opening-up policy was initiated. In short, the policy began a process by which the economy became liberalized. The Guangdong Province and the Pearl River Delta (PRD) region were essentially the leaders in this dramatic turn in policy, providing areas or preferential treatment to businesses. As a result, a number of Hong Kong enterprises started to move their production facilities to the PRD region. It was then that Simon's gut feeling told him to move.

When building the factory, his initial idea was to locate the factory in Shenzhen, which was close to Hong Kong and convenient to manage. The Shenzhen

❖ *Simon in his office at Wing Kut Industrial Building, Castle Peak Road, Cheung Sha Wan*

government welcomed investments from Hong Kong and granted him around 4.7 hectares of land to build the factory. All the paperwork was ready, the only thing needed was his signature for the construction to commence. Just then, his father introduced Simon to Mr. Zheng Jintao, the then Chief of Dongguan County. At that time Dongguan was lagging in development and in desperate need of industrial investment. Zheng wanted to persuade him to set up a joint venture with the Dongguan County Government by building a factory to manufacture labels in Dongguan. Simon was moved by Zheng's pleas, whose words evoked for Simon a deep affection for his hometown. Simon's idea was crystal clear: it was good to contribute to the development of his hometown. At the end of 1983, a Sino-Hong Kong joint venture, the Dong Ying Computer Label and Embroidery Limited was founded.

As Dongguan was actively promoting industrial development, its government provided generous resources and support for the joint venture. Although Simon had some savings, his strength had rested not in his financial resources but in his familiarity with the industry, his knowledge of the market, and his ability to develop business — an expertise that the local government was plainly lacking. His idea of using Hong Kong's trade network, the Mainland's construction resources, and the

❖ *In early 1984, Simon visited label factories in the UK to prepare for the joint venture.*

advanced production technology from abroad to promote the development of the joint venture was, to say the least, very forward-thinking. Zheng greatly appreciated this forward-thinking young man.

"The most important component of a loom is the jacquard. And the most advanced one is the electronic air jet jacquard made in the UK. Vaupel in Germany and Muller in Switzerland produce the most famous brands of looms."

Simon was very familiar with the looms. When he visited the UK and Germany in 1984 to prepare for the joint venture, he was inspired by the fact that the label factories there were not labour-intensive, relying instead on advanced equipment to create an assembly line.

Talking about his early experience of travelling abroad, Simon still recalled in vivid detail: "The stamp on the permit is black for those who have lived in Hong Kong for seven years and have obtained the right of permanent residence, while it is green for those who have not lived there for seven years. At that time, I had been in Hong Kong for only four years, and I had a green stamp. I had to go through a lot of trouble to get a visa. The embassy did not want to grant me a visa. I had to ask the bank to offer a guarantee to get the visa."

In the early days of the venture, this was, however, a relatively minor obstacle next to the most imposing challenge: how to get the machines running.

He ordered a batch of looms from the UK, the first of its kind in the label industry in Hong Kong and the Mainland. The computerized machines were very fast and sophisticated, and, as no one had seen anything like it before, the workers had to feel their way across the river and try things out as they went along. Despite the presence of British engineers on site to provide technical assistance, the local workers in Dongguan were unable to employ the technology. For the first six months, the factory was not able to put the machines in operation. Additionally, there were several technical issues, and each day, Simon had to travel between Hong Kong and Dongguan.

As soon as he finished his work in Hong Kong, he rushed back to Dongguan to meet with workers in the factory to solve technical problems. He did not sleep much during those days: either staying in the workshop all night or thinking about the operation of the machines. In addition to the technical difficulties, his partners in the joint venture began to doubt the viability of the factory when they saw that the factory had not been able to reach expected production levels for the first six months. Simon began to feel the pressure.

Fortunately, Mr. Zheng, the Chief of Dongguan County, also heading the joint venture, was very supportive. The technicians made every effort to solve the issues. In the fall of 1984, the machine finally became fully operational. Simon had passed the first hurdle. In December of the same year, the Chinese and British governments signed the Sino-British Joint Declaration, affirming that Hong Kong would be returned to the Chinese sovereignty in 1997. The agreement marked a new milestone in the economic interdependence between Hong Kong and the Mainland.

Simon understood the importance of having the right machinery and the right people. In the 1980s, the label industry was still relatively new in the Mainland. Professional staff was almost a rarity, and finding those with a background in the business was a near impossibility. However, for Simon, it didn't matter whether the staff had a related background. In fact, it might be even better if they didn't because they could bring a new perspective.

"I want to hire people who are quick learners with a good foundation in mathematics and science to pick up technical knowledge fast. I want to rely on new ideas, new technologies and new models to promote production, rather than intensive labour."

He recruited quite several teachers from local schools, including his math teacher Sun Jingxian, and arranged for them to receive professional training in textile technology at a silk weaving factory in Beijing to develop a pool of technical experts.

"The factory was huge, with a whole floor full of looms, all of which were of the most advanced type, costing almost a million HK dollars each. It had only been a little over two years since he started looking for orders, but Simon had already built up his business to a very large scale. Since then, I felt that this man was quite remarkable. He had the charisma and ambition and would definitely succeed," recalled Raymond Leung.

In October 1985, Simon, still holding a green stamp permit, obtained a visa again through a guarantee by a bank via his father's connection, and was able to visit the label industry in the US. In many of his subsequent interviews, he mentioned this trip to the US and a Jewish man called Rollin Sontag.

Sontag was quite literally born into the labelling business. His father acquired a labelling company, Capitol Label, on the very day of his son's birth. Such an event seemed to foretell an indissoluble bond between Sontag's life and the label industry. Soon after joining the family business in 1948, Sontag felt the need to expand the business. He wanted to create a synergy — an idea that his father endorsed. In the same year, they acquired American Silk Label (ASL), a long-established American label company.

Founded in 1875, ASL had been in the industry for over 70 years and was certainly a long-standing name in the American label industry. After the acquisition and under the careful management of Sontag and his father, ASL quickly expanded production from a few million a year to upwards a billion labels in 1984 with production facilities in both New York State and Philadelphia. At the time when Simon met him, Sontag had been the President of ASL for 20 years. At his estimate, the US label market at the time totalled approximately 70 million US dollars. ASL accounted for 20% of that market, making it a significant player in the industry. At the time of its acquisition by Sontag, ASL's sales revenue was only 350,000 USD while now the total market share had been 14 million USD, a 40 times growth.

*The New York Times* on 9 September 1984 ran an interview on page 11, entitled "IF THEY CAN MAKE IT, HE CAN LABEL IT". The interview read: "Although he (Rollin Sontag) is reluctant to name his customers, he admits that almost every major retailer and garment manufacturer is a customer of ASL."

On his visit to the United States, Simon was hosted by Sontag, the exact one of the stars of the US label industry.

Sontag was a visionary within the label industry. He advocated for innovations, including upgrading, and developing new production equipment as well as computerizing much of the production process: a development which, thereby, enabled quality control. He even equipped a cathode ray tube (CRT) terminal and designed a management platform that could track business orders. Even though his factories were in different states, the standardized production management model was able to ensure consistent quality. Sontag took Simon to visit his factory in the state of New York, which put the one-stop service model into practice. The model employed technology to maintain consistent quality, combined with creativity and design. Simon recognized the model as eminently usable and found the blend of quality control and innovation to be inspirational. Most importantly, the young man recognized Sontag as a forward-looking and innovative entrepreneurial spirit.

The year of Simon's visit, Sontag had also been elected the Chairman of the Board of Directors of North Shore University Hospital. The Hospital served as a teaching unit for Cornell University Medical Centre. Sontag told Simon that after becoming Chairman, he used to wander around the hospital where people would

❖ *A visit to ASL in 1985. From left: Suen Shing, Rollin Sontag and Simon*

❖ *Simon (first from left) visited the ASL production workshop to learn about the label production process.*

poke their heads out of the door and say to him, "Hey, I know you!" It was very interesting and very satisfying to him. The story moved Simon deeply. From simply wanting to run a good factory, he began to have a new understanding of the social value of business: from the society and for the society.

Despite their age difference, Simon and Sontag became very good friends. At that time Sontag was near his sixties. He already had five children from two marriages, but none of them were willing to carry on his legacy. For Sontag, who had long understood the need to provide a one-stop service with consistent quality, it was a long-held dream to expand his label business globally and become a worldwide leader in the industry. Simon's willingness to dive headlong into the label industry deeply impressed Sontag who readily shared his experiences and ideas while making every effort to entertain Simon who was new to the US. He took Simon to and from New York State and Philadelphia on his private jet to visit the factories in both places, bringing Simon, as well, to a Broadway show. Although Simon had no interest in the theatre and fell asleep in the middle of the performance, only awakening to the applause of the audience, he was touched by the gesture. At the end of the trip, Sontag also hosted a huge farewell dinner for Simon.

Just before boarding the plane, the ambitious young man was photographed on the hardstand. In the photo, he wore black-framed glasses, a dark khaki check suit and a light khaki scarf in his left pocket — a very smart-looking colour scheme. He was not yet thirty years old, his eyes were sophisticated but not worldly, and his smile youthful but not confused. The young man had a secret ambition to realize Sontag's unfulfilled dream of becoming the world's number one label company.

To commemorate the friendship with Sontag and his role as an inspirational figure, in 1997, Simon changed the name of his business to SML Group with a similar naming style.

After returning from his visit to the US, Simon set about to make a big move. To accelerate the development of his Hong Kong and international business, Simon and his wife set up Dong Ying (Hong Kong) Computer Label Development Company Limited in Hong Kong, equipped with basic production capability, which located in the Fu Cheung Centre at Wong Chuk Yeung Street in Fo Tan.

"The reception area was so grand and modern that many people passing by mistook it for the office of a bank," a former colleague recalled. Simon laughed, "It's true. Even the sofas were bought from Germany." For Simon, who has always been frugal and disciplined, his willingness to invest heavily in the renovation was not a "vanity project". It was a result of his overseas visit. The label industry was closely related to the garment industry. A company's understanding of trends and fashion had an invisible impact on its business development thus building its image as a modern enterprise was important. Simon was able to apply his macro knowledge of the industry to micro practices: from picking up an old desk to opening Hap Seng Company in Fu Tor Loy Shopping Centre, to equipping Dong Ying in the Fu Cheong Centre with German sofas. He had been able to make quick adjustments in response to the changing needs of the business and paid attention to details, even at the beginning stage of his business.

It was also in the same year that this young founder won Esprit as a major customer, an event pivotal to the development of the SML Group as an industry leader.

"A label can be as small as 20mm long and 10mm wide, but we are not just producing a label. We are passing on and shaping the values of a brand."

From the company's founding, Simon had a clear vision. The label was not merely an object. It was a vehicle to express the identity of a company, explaining why it was so difficult to gain approval from a major brand — in particular, from a brand with a well-established reputation.

In Simon's view, the quality, rather than the quantity of customers, is more important. If you can gain approval from a quality client, you will definitely heighten your reputation in the market. This enhanced reputation can be leveraged to promote the busines growth. This was, in other words, why, early on, Simon set his sights on the German apparel brand and why his determination yielded success.

Esprit had stringent requirements on its labels from pattern design to the yarns. Its requirements on the quality and delivery were even more demanding. After strenuous efforts, Simon managed to get shortlisted with his samples, but still a long way to success. Yet Simon took it as normal. In his view, every endeavour has its difficulties. The only response to an apparent rejection is to solve step by step each problem: to keep improving and progressing steadily. The whole Esprit team in Hong Kong was very impressed by this hard-working and resourceful young man.

From sporadic orders at the initial stage to 1985 when the company acquired approval for all of Esprit's core product lines and became its major label supplier spanned only two years. Simon's judgement was spot on: Esprit's sales went through the roof and remained among the world's most sought-after apparel brands for the next two decades. Just as he predicted, his business rose dramatically. "At that time,

❖ *Simon (front row, right) visited Esprit's headquarters in Germany several times and met with the then Chairman Heinz Krogner (front row, left).*

it was so profitable that producing woven labels was like printing banknotes. Even the purchase of expensive looms specifically for a purple label paid off in less than six months."

The good times didn't last forever. Since 2018, Esprit had underperformed. Although the company's main business had long shifted to RFID intelligent label, Simon was still grateful to Esprit for their early partnership: "We grew up with Esprit. Without Esprit, SML would not have found its place as a global leader."

SML Group continued to expand its operations, with the joint venture up and running in Dongguan. On 13 December 1986, *Nanfang Daily*, the Guangdong provincial newspaper, featured on its front page a photograph of women in uniforms standing beside a loom with the following description:

> *The Dong Ying Computer Label and Embroidery Limited became the first of its kind in China in the 1980s, to have a state-of-the-art computerized electronic jacquard machines. The factory used the quality nylon yarns to manufacture brand labels for various ready-made garments and was, thereby, able to market their products globally.*

To understand the achievement, one should recognize the state of label industry in China in the early 80s. Some factories were still using old machinery with outdated technology. Simon recognized the vacuum and, unlike his competition, used advanced machinery, provided design services, and implemented computerized management systems. Two years after its establishment, the joint venture was already making profits and generating economic value. In a brief time, the Dong Ying factory became famous as an exemplar of innovation and quality assurance for the manufacturing industry throughout Guangdong Province.

In January 1988, Dongguan was officially upgraded to a prefecture-level city, one of the four prefecture-level cities in China that did not have districts at the time and was directly under the jurisdiction of Guangdong Province. Mr. Zheng Jintao, the Mayor of the city cum Chairperson of the joint venture at the time, was still very supportive of Simon's endeavour. In the same year, Mr. Li Peng, the then Acting Premier of the State Council, visited Dongguan and made a special tour of the Dong

❖ *Photo on the front page of Nanfang Daily on 13 December 1986*

❖ *Li Peng (right), the then Acting Premier of the State Council, visited the Dong Ying factory in 1988, with Simon on the left and Zhang Jianming — manager of the factory — in the middle.*

Ying factory which was regarded to be a role model for advanced manufacturing.

It was precisely at this point, Simon decided it was time to "go global."

He first presented the idea at a regular meeting with his joint venture partners. The partners strongly opposed the proposal. At that time, the Dong Ying factory was a household name in Dongguan, and anyone who had some social connections wanted to seek employment there. The factory was considered a huge success, and it seemed to many that it was only sensible for the enterprise to take advantage of such success and make profits.

But Simon didn't see it that way.

He noticed that the success of Dong Ying had led to an increase in outside investment. Other emerging players in the industry were beginning to follow a similar business model, relying on original equipment manufacturer (OEM) to take advantage of low labour costs and favourable foreign investment practices; thus far, the model had achieved a satisfactory level of profitability. Such an increase in competition would have adverse effects on price. The companies would simply bid the price down, eating into profit margins. Short-term business needs might be satisfied, but to have long-term growth, the company had to reach another level.

Moreover, Simon had larger ambitions. He aspired the company to be a global leader and not simply to be regional label manufacturer. His next proposed step was to go beyond the low value-added processing and production model, establishing direct partnership with international brands in order to expand his businesses globally.

At that time, Simon was just over 30 years old and was already very forward-looking.

Quite a number of members of the management team of the joint venture were government officials. While they admired this young man's ambitions and ideas, when he talked about globalization, they could not agree with Simon. For these officials, globalization sounded an illusory concept. Their attitude was that

❖ *Liu Wen (left), the then Vice Mayor of Dongguan, was a member of the joint venture team and an admirer of Simon (right).*

the existing model was effective and should be pursued in a pragmatic manner to maximize profitability.

"I held 40 percent of the venture. However, the government held 60 percent, so technically, the government was my boss. But we had a serious disagreement about the position of Dong Ying." After a heated debate, Simon proposed to relinquish his role in the joint venture. He was determined to follow his dream and to realize his pursuit of becoming number one in the world.

The joint venture team sitting across the table was stunned to hear that he was so serious and determined. Of course, they didn't want to lose this talented young man and knew well what it meant if he started a new business. They tried their best to persuade him. But Simon was not swayed. He decided that even if his idea finally met their approval, there would likely be a great deal of resistance to its implementation. Since they had different views, it would be difficult to head in the same direction. It was better to make a quick decision than to waste time with tied hands. As the general manager of the joint venture who was responsible for the profit and loss, he felt that his quitting did justice to Mayor Zheng who had given him the opportunity.

The manager of the factory, Zhang Jianming, was born and grew up in Dongguan. After graduating from college, with an economics major he was assigned to the Dongguan Economic Development Company. When the Dongguan government and Simon established the joint venture, he was appointed by the government to be the factory manager. He was at a similar age and got along well with Simon. The day after Simon decided to quit, Zhang walked straight into his office, closed the door, pulled a chair beside him, sat down and said to him: "You will definitely lose by going your separate way. What about capital? What about government support? Where are the resources? You're going to lose everything! It's all bad for you!"

"I know he meant well; he was afraid I was being overly impulsive." Thirty years later, Simon recalled the episode. His office faced the sea and overlooked the old Kai Tak Airport, with a view of the Convention and Exhibition Centre and the Golden Bauhinia Square.

"Joint ventures are like marriages: if they work well together, they will be happy, but if they don't, they won't be sweet. Success relies on my own capability. I believe I have the ability and wisdom to succeed."

Zhang was unable to convince Simon to remain with the company. Later, Zhang left for San Francisco to settle in the US. Whenever Simon visits the US, he will make a detour to San Francisco to visit Zhang. However, the last time they met was not in America but when Zhang returned to his hometown in China to visit his family. Zhang recalled, it was in May 2019, and Simon made special trip from Hong Kong to Dongguan to dine with him.

"I talked with him a lot. But in the end, he told me that I was wrong and he had the good judgement. When a couple gets divorced, what often matters most is how one is on one's own." Years later, Zhang remembers most vividly this analogy.

Simon started with a clean slate. He left the factory to his partner and only requested three colleagues: his former maths teacher, Sun Jingxian, and two others, Ye Songgen and Li Hanwei. The latter he recruited personally from Dongguan and sent to study at the Beijing Textile Institute. His request was of course approved. So, with a team of just three people, Simon dedicated his energy and drive to realizing his dream becoming "number one in the world".

Almost immediately, an obstacle occurred. It was 1989, and due to general uncertainties in the economic and political climate, the banks demanded the immediate repayment of the debt from the Dong Ying joint venture. As a shareholder from Hong Kong, Simon had signed as the guarantor of the loan. The amount was no doubt astronomical to the young entrepreneur.

"I definitely did not have the money to pay the debt off in one go," Simon said.

He had the audacity to go and talk with Poon, the bank manager. Simon is not a blind optimist. In his view, when encountering difficulties, one must possess three qualities: the courage to face the difficulties, the determination to solve them, and most importantly, self-confidence. He believed that difficult times were the time to turn pressure into motivation, crisis into opportunity, and bad things into good.

Simon's approach was, as usual, straightforward: "Honestly, I am not able to repay the debt at the moment. If I have to be held legally responsible for this, I have nothing to say because I did guarantee the loan for the joint venture. But if the bank will give me some time and allow me to pay in installments, I will keep my promise and, without fail, pay the sum back."

Perhaps, it was Simon's honesty, courage, or his determination and self-confidence that made Poon agree to his proposal and allowed the debt to be repaid in installments. From the bank's perspective, what mattered most was that the loan could be paid off.

It was not until 1995 that the note was repaid in full, but Poon's trust was rewarded.

Many people were puzzled over Simon's decision to leave the joint venture — especially when the banks called for an immediate repayment of his note. The joint venture, after all, seemed a secure and promising business and the goal to globalize his business appeared at best to be a dream. But Simon was not convinced. To "lose now" is to "win in the future", and to be bold enough to lose is to be bold enough to bear the consequences of losing.

It is not unusual to have the will to win, but it is challenging to have the courage to lose. To be willing to give up immediate, concrete, and vested interests in pursuit of seemingly illusory and unattainable long-term values requires extraordinary courage. Simon, only in his early thirties, already had the profound understanding of an important principle in business and life. It was thanks to his courage to let go and even to accept the consequences of that loss to realize his dreams that he would eventually become a global leader in the labelling industry.

After nearly a decade of painstaking efforts, the young entrepreneur had, with a back-to-basics mentality, embarked on an ambitious journey in line with his vision of "becoming number one in the world" — an entrepreneurial journey he described as "hard, but with joy in the midst of pain".

❖ *Simon in his office at Dong Hing, 1994*

CHAPTER

# III

# INNOVATION

/

## GAINING A FOOTHOLD

# Channel All Strengths to Tackle One Point

In May 2018, Simon went back to Shangsha to visit the retired "Uncle Jing", a respected name for Sun Jingxian, and to have dinner at the factory canteen. As the car drove into the factory, he looked at the fiery blossoms on the trees on the road and said, "The flowers are blooming so well this year!" He had handed over the management of the factory to his subordinates years ago; but inadvertently revealing his deep affection for the factory that he had built with own hands.

Simon named the factory in his hometown Dong Hing. "My grandfather's name was Da Xing, so I named the factory after him." The character, "Xing" in Chinese, classically suggests "to rise up" as well as "to flourish". For Simon, the factory combined both meanings, representing his hopes and intentions for the factory.

Simon always has an emotive attachment to his hometown. When selecting the site for this sole proprietorship, he could have set up the facility in Shenzhen, but he decided to accept the local official's invitation and returned to Chang'an with a team of only three staff members to build a factory in Henglong, north of Shangsha Village.

Henglong faces the south and is flanked on three sides by mountains. At the time of the factory's construction, the village was only connected to the outside world by a winding road. A spring trailed down the eastern slope of a hill adjacent to a small reservoir. According to the seniors in the village, Henglong was a treasure land gifted with the best Feng-shui.

Nevertheless, Henglong did not appear to be the most advantageous location for a factory. It was not supplied with electricity. At night, the woods were tranquil and dark. But Simon felt a light beaming within his heart.

He often spoke of the image of light. Upon his first arrival in Hong Kong, he looked at the lights and stars of the city at night and experienced a sense of anticipation in his heart. This hope echoed with the lyrics: "Let every beam of light shine, and shine with hope", written by Cheng Kwok Kong, the lyricist for the theme song "This Is Our Home" of RTHK's "Dreams of Hong Kong" project in 1990.

When they met for the first time 27 years later, Simon, a former secondary school teacher, told Cheng, a former primary school teacher, how his lyrics had accurately captured his feelings when he founded Dong Hing. Simon invited Cheng to his 60$^{th}$ birthday celebration, where the colleagues of Dong Hing sang "This Is

Our Home". Standing on a barren field of red mud at that time, Simon was convinced that the Dong Hing factory would lead the way in revolutionizing the label-making industry.

The first days did not, however, seem to predict success.

Sun Heyi and Sun Jingxian, aka Uncle He and Uncle Jing respectively, both agreed the first few years were an ordeal when the factory was being constructed — then, working at Dong Hing until their retirement several decades later.

"Before the factory came here, it was a wasteland. There was nothing here," Uncle Jing points to the modern factory. Nearly 80 years old, he remains energetic, riding his electric bike everywhere.

He still remembers the early days vividly:

"In the beginning, there was only one factory building and eight employees, who all lived at the factory. Production facility was on one side, the living area on the other. Outside the factory, there were barren hills. When it was time for meals, we went to the hills to pick firewood and cook in the wild. We all grew up in the rural areas and were used to such poor conditions. No one complained about it."

"There was a shortage of everything: money, electricity and manpower, so we had to do it all by ourselves," Uncle He recalls. He is ten years younger than Uncle Jing and is from Shangsha Village where he used to drive a tractor for the production brigade.

"Simon was very determined, so determined, in fact, that he would push through any difficulty and do anything to get Dong Hing built," Uncle Jing said.

For Simon, though, the challenges were not unexpected. "Everything," he said, "is difficult at the beginning." Perhaps from the moment he left the Dong Ying joint venture, he realized already that the road ahead was not an easy one.

"Nowadays when building a factory, the money must be in place once the project is approved. The construction must be carried out according to a pre-approved schedule, and the factory must be fully completed before being put into use. But in the past, construction could be started before the project was fully funded. Later on, any surplus cash from the company in Hong Kong would be sent to Uncle Lam. Be it $300,000 or $500,000: as long as the company had the surplus, we sent it over. All that mattered was that the project wasn't interrupted." He even made a funny gesture of clutching a pile of money and handing it over as he talked about the past days of building the factory. His audience laughed, and Simon laughed along with them.

In addition of repaying a huge debt of the joint venture, he had to raise funds for the construction of Dong Hing and developed his company in Hong Kong at the same time. There was a great deal of weight on Simon's shoulders. It is not difficult to imagine the pressures he must have felt at the time. Only his extraordinary vision and perseverance helped him overcome his ordeals and become a success.

Suen Lam Fat, who everyone called Uncle Lam, was born in Shangsha before immigrating in his early year to Hong Kong, returning to the village after working in the construction industry for many years. He was responsible for the planning and design of the Dong Hing factory. Simon gave him full authority over the whole project and only made two requests: firstly, the factory shouldn't adversely impact the surrounding environment. In fact, Simon requested that more trees be planted. As such, Uncle Lam planted many trees in the Dong Hing factory: in addition to the banyan, osmanthus and oleander trees, he also planted fruit trees, including lychee, mango and longan. These trees were so sturdy that they were not afraid of being blown over on typhoon days in the south. In good harvest years, around June, the canteen would have baskets of lychees and longans to share with colleagues. Secondly, Simon requested that there should be a living area with supporting facilities such as dormitories, a canteen, a sportsground, a library and a medical clinic. Uncle Lam followed these two general directions in formulating his design. The planning model for the first factory is still displayed in the reception hall of Dong Hing: on the top left corner of the model are four words — "Green water, green hills".

"Nowadays we talk about corporate social responsibility, but back then, the idea was simple: we must not destroy the natural environment; we must attach importance to the working and living environment of our employees and give them a sense of belonging. Simon's approach was to "see the essence through the phenomena". He understood that talent was the greatest asset. At a time when there was a huge construction boom, it was rare to find an entrepreneur who was willing to spend money on both staff welfare and had the conciousness of building an eco-friendly environment. His holistic view and farsighted planning has proven time and time again to serve Simon's business well.

"Basically, a new factory building was built every year since the start of construction, and we continued to build new factory areas. We built staff dormitories and canteens as soon as the factory building was completed, and we had someone to take care of the catering." Sun Heyi, who was in charge of administration and finance, laughed that the "corporate structure" at that time was "small but complete".

❖ *Designed in accordance with Simon's ecological concept, the Dong Hing factory locates in a beautiful environment surrounded by green mountains and water.*

"Uncle Jing was in charge of production, while Uncle Lam was in charge of factory construction, and Simon was the chairman, overseeing the whole business." The clear division of labour and responsibilities led to a fast yet steady development of Dong Hing. What had once been an impoverished and sleepy rural community became a bustling workplace nestling in a vibrant natural environment.

Meanwhile, Simon had to travel frequently between Hong Kong and Chang'an. At that time, there was only one National Highway 107, so it took at least three to four hours, and sometimes even six to seven hours, to cross the border from Lo Wu, followed by taking a bus to Shangsha. Many a time, he arrived in Dong Hing late at night. However, as soon as he got to the factory, he summoned colleagues for meeting. Many colleagues who joined Dong Hing in the early days of the founding of the factory still remembered being notified of meetings at 9 PM or 10 PM because Simon had only just arrived at the factory. The road between the dormitory and the factory had not yet been paved, so they had to tread carefully in the dark through a muddy path. A flashlight was "standard equipment" for protection against snakes. When talking about the late-night strolls, several long-time employees burst out laughing.

The entire team emulated Simon's enthusiasm and drive. They spent many hours at the factory. They were so busy that they couldn't even tell what day of the week it was. The equipment was kept running at night, so several supervisors took the initiative to patrol the workshop to ensure normal operation. Later, when the business was on track, Uncle Jing, the then deputy general manager, set up a bed in the porter's dormitory so that in case anything happened in the factory, he could respond at the first opportunity.

To build a good factory, a good team must be built first. Modern management theory stresses that business should be "people-oriented". Choosing the right people can yield twice the result with half the effort; choosing the wrong people, however, will get half the results with twice the effort. Simon's idea was simple and practical: "To gather wealth, first gather talent". Back in the Dong Ying joint venture, many people sought employment there through their social connections because of its good performance. Simon insisted on employing talented people, which resulted in some internal friction. Now in his solely-owned enterprise, he embraced all kinds of talent to build a strong team for Dong Hing. In his view, different people have different strengths and weaknesses, but the key is to give full play to their strengths. The appointment of Sun Jingxian was a good example of how Simon was capable of using people well.

Sun Jingxian had a mathematics background. So, the first time he was put in charge of production, everyone was puzzled. In the early days of the factory, although the volume of orders was more than sufficient, the profit level was not satisfactory, and the productivity was not high. Sun Jingxian reformed the two-shift system to a three-shift system, with lunch shifts to reduce the downtime for machines. As a result, in a very short period of time, the productivity grew by 80 percent. He also developed a set of cost calculation formulas, which he derived from the complex label weaving process in order to effectively control production costs and improve productivity. He changed the assessment of workers' performance from a strictly hourly rate to piecework, advocated the concepts of "efficient and inefficient operating of machines". His system rewarded efficiency as opposed to time spent on the task, thereby, raising productivity. As a former teacher, he also set up a training school with Simon's support to teach the know-how of label weaving, requiring the management, sales staff, that each member of the team become familiar with all aspects of the product, production and costs. They were required to attend two classes a week and take an examination at the end of the course. If that staff failed, the course must be retaken. This training scheme lay a solid foundation for developing the talent pool.

Management does not necessarily need a management degree. Sun Jingxian had the facility to apply his background in mathematics to analyze and discover solutions for problems in a logical and focused manner. This is what Simon meant when he used the phrase, "the crux of fresh perspective."

The following words in the song of Dong Hing capture the spirit of the factory: "In the East raises a new star of label weaving with state-of-the-art technology on a large scale. We are committed to good service, quality assurance, and computerized management, aiming to become a world trademark enterprise." Sun Jingxian wrote the lyrics in one night. Simon had often preached these words, a message that had found its way deep into the heart of each member of the factory team, so did Sun Jingxian.

Under his steady leadership, the team did not fear enduring hardship and demonstrated a continued drive and commitment; Dong Hing was, thus, able to enjoy rapid growth. In 1992, Simon commissioned a sculpture of a bull; then placed by the entrance to the Dong Hing factory. Its hooves were dancing in mid-air. "Prosperity comes from diligence, just as harvest comes from cattle's hard work":

❖ *The bull sculpture at the entrance of the Dong Hing factory is a testament to the growth of Dong Hing.*

the sculpture, meant to inspire the Dong Hing employees to keep alive the spirit of the "pioneering cattle".

Dong Hing was built during a period of favourable conditions. The label business was complementary to garment production and its growth became linked to the burgeoning rise of the garment industry. Hong Kong's garment industry had relocated on a large scale to the PRD region. Many major customers had begun to come to Dong Hing: their orders sparking the development. The business was booming and a financial surplus was beginning to emerge. However, Simon, with his hard-working team, invested all the profits in expanding the production capacity of the factory, purchasing advanced equipment and improving staff welfare. It took only a few years for Dong Hing to expand from only one workshop with four broad looms and four shuttle looms to four workshops, subdivided into a broad looms workshop, a shuttle looms workshop, a needle looms workshop and a finishing workshop. The factory had on site over 200 machines. Observing the company's exponential development, the banks offered loans, funds that supported Dong Hing's further growth.

In the 1990s, with the reform and opening up of the country, those who made a fortune were the first to buy the big items from tape recorders, washing machines, colour TVs, telephones to air conditioners. But that was not true of Simon who was only interested in buying looms. The more advanced the technology, the more he became interested. Frequently, he attended various exhibitions to keep abreast of the latest industry developments.

In 1993, at the International Textile Fair in Beijing, a well-known German textile equipment manufacturer, Vaupel, displayed a high-speed air-jet loom, the first of its kind in China. Simon walked around the machine three times. He simply did not want to leave. Before the exhibition was over, he directly approached the representative of Vaupel at the site and asked whether they could sell the exhibited machine to him.

A loom cost a great deal — over one million RMB. It was definitely a significant investment. If the money had been used to buy land, it could buy around 0.7 hectares. The value of that investment would have multiplied by 100 times in 20 years' time, and his fortune today might have been astronomical. To this calculation, Simon's answer was always the same: "It's good enough to just do one thing well." As an entrepreneur, he is highly dedicated, driven by an almost maniacal insistence on achieving his ultimate goal.

The representative of Vaupel contacted his headquarters to ask whether the exhibited machine could be sold directly. Hans Vaupel, the General Manager of Vaupel answered the phone:

"It was Simon Suen, right?"

When Vaupel heard that a label manufacturer from Dongguan wanted to buy the most advanced looms, his first instinct was Simon.

Simon gave this English name to himself, which is close to the pronunciation of his Chinese name. In the 1980s when went overseas to visit the label industry, he felt the need to have a catchy, easy-to-remember English name — paid a great deal of attention to the long-term impact of details. The name, Simon Suen, indeed had, since then, become well-known in the global label industry.

"How did you know?" The representative was full of surprise over the phone.

He did not realize that the two gentlemen had already met in 1987 during Simon's visit to the headquarters of Vaupel Group at Wuppertal, Germany. Vaupel had received him then. One year after, Vaupel also visited Dong Ying at Fo Tan, Hong Kong.

Vaupel looked typical for a German: he was tall and strong with a wide face,

❖ *Dong Hing took the lead to use air-jet loom in the 1980s.*

❖ *Vaupel (first from right) visited Dong Ying in 1998 and Simon (middle).*

sharp jawline and well-defined cheekbones. He had curly brown hair and wore rimless eyeglasses with golden frames. Back in the 1980s and 1990s when woven label production still relied on traditional techniques and was yet to be modernized, Simon's vision and effort to import advanced technology and equipment impressed Vaupel deeply. He admired this Chinese man who, though of average build, was very ambitious. He thought of Simon not only as a business partner but also as a friend. For Simon, whenever he attended an exhibition in Germany, he would drop by Vaupel's house at Wuppertal. Vaupel would bring him to Neue Welt, a famous German restaurant for knuckles.

After his visit to the exhibition, Simon bought the broad loom and had the machine installed at Dong Hing. The loom was used for 20 years and has been preserved in the factory until today, which denoted the commitment of Dong Hing to assume a lead position in revolutionizing the label industry.

From the end of 1980s, the model of "front shop, back factory" became prevalent in Guangdong and Hong Kong. Renowned as a hub of international trade, Hong Kong was the "front shop", taking overseas orders while managing international marketing and sales. With vast land and cheap labour, the Pearl River Delta served as the factory where the products were produced. Dong Hing and Dong Ying operated together according to this model and developed rapidly in Southern China. They soon found their feet in the label industry. By 1994, Dong Hing achieved a large scale and the debt for the joint venture was almost paid off. It was then that the young Simon officially put forward the agenda of business globalization. Actually, in the decade from 1985 to 1994, Simon had never stopped pondering the same question: how to and what was necessary to globalize his business?

After many international business trips and attending industry exhibitions, Simon formed his own judgement about the future development. Based on his observation, to form a global business network, direct connection with clients having a global sales network is a must, which can only be established through systematical connection of electronic data interchange (EDI), rather than relying on agencies. To initiate this step, the company needed to launch an Enterprise Resource Planning (ERP) system first, to manage orders digitally.

Simon had a typical goal-oriented mindset: through careful and step-by-step analysis, he could turn abstract vision into concrete action plans.

His first step was to initiate the ERP system in Dong Hing. There were no global customers per se, so at first, this step had no real impact on the bottom line. The factory still relied on agencies to obtain orders. But this move implies the future.

Once again, unlike other producers who chose to adhere to the familiar and as of yet, profitable products, Simon invested heavily in launching the ERP system. Simon did not feel the need to defend his strategy which was no doubt seen by many as "a waste of money". Twenty years later when he became a successful entrepreneur, he was often invited to share his business philosophy with young people. His most important advice was "to look forward".

Simon said: "A workman must first sharpen his tools if he wants to do his work well; then hide his tool and bide his time. With ERP and EDI in place, we were well prepared for opportunities."

Dong Hing launched its ERP system when most label factories still managed their orders manually with paper, pen and tied knots. They recorded orders on paper and hung the paper with tied knots. Simon's philosophy was to "let the professional do professional things", and he spent a lot of money hiring a software company to develop the ERP system.

"I met Simon for the first time in the spring of 1994," said Roger Chau, who had been the IT director of the company. Wearing spectacles, looking gentle and refined,

Roger had taught at a university before he joined the system software company, where he was assigned to carry out SML's ERP development project.

"In the 1990s, few garment factories used the ERP system, not to mention label factories. Many of my company's clients were large-scale electronic enterprises. It would take a lot of money, resources and personnel to launch the system. Few companies would have a try unless they had developed to a reasonable scale."

Roger was unsure about the practicality of the plan until he met Simon. However, after talking with Simon, Roger's suspicion was transformed into admiration.

"I didn't expect that a man at such a young age could manage a company of several hundred people; I was even more surprised that he not only realized the importance of IT but also knew a great deal about the technology."

Roger commented on the RS6000 small server he saw at the company when he visited Simon for the first time. "It's not an ordinary server," said Roger.

This server was designed and produced by IBM in 1990. Although a small computer server, it was not a general one at all. It belonged to the same RS6000 series as Deep Blue, which beat the world chess grandmaster Kasparov in 1997. RS6000 servers were expensive for their excellent performance and quick processing speed. Roger was so surprised to find the latest server in the IT industry at a factory producing labels, which further convinced him of Simon's determination to apply technology in production.

Later, Roger accepted Simon's offer and became the company's IT director not long after he started his own business. "How foolish I would be if I refused to serve a wise master," Roger said. His later experience at SML proved that it was definitely the right choice.

"As we're developing the internal ERP system and the external EDI system at the same time, our IT department needed to hire more hands as soon as possible," Roger recalled. More personnel meant higher operation costs, but more input did not necessarily ensure an immediate positive effect on the business. Roger said, "I was worried that Simon might not agree with me, but he was very supportive. He told me IT meant high productivity and would be the driving force for development." Simon's understanding of the importance of IT greatly impressed Roger who was himself an IT specialist.

Roger continued, "Around 2000, Simon asked the IT department to follow the RFID technology and trends in the market. He also asked us to absorb more specialists in the field to prepare for the future." Simon's decision proved prescient. Several years later, apparel retailing and other areas began to apply RFID technology. The company was able to find its footing in a rapidly evolving market. Roger owed the success to Simon's prescience and to his emphasis on talents, saying, "This showed his vision and ambition."

It was also in 1994 that Yang Yong, the pattern design manager of Dong Hing, met Simon for the first time. Yang came from Hubei Province and began a career at Dong Hing, lasting for over 26 years. He graduated from Wuhan Textile Engineering Institute, the predecessor of Wuhan Textile University. After graduation, he worked at Caidian Towel Factory for nine years and took charge of towel production techniques. "The job was fine though not exciting," Yang said. "However, I had heard there were many opportunities in the South and was considering whether to take a change or not."

In March 1994, unbeknownst to his family, Yang went to the Wuchang Labour Bureau. Yang said, "I was not sure whether I could get the job. Then, I met Simon. My first impression was that he was very young and not at all outgoing. He wore a very ordinary T-shirt and didn't look at all like a big boss from the South." After reading Yang's resume, Simon had only one question: "When can you come to work?"

Yang did not expect the apparently reticent young boss would be so straightforward and decisive. The traits, impressed him a great deal.

After the reform and opening up in 1978, Guangdong, a coastal province in Southern China, became the top destination for immigrant workers from other provinces in China. In the 1990s, Guangdong accelerated its development and became a "land of gold" where people all around the country flocked to make a living. According to Sun Heyi, the then administrative officer of the company, there was always a long line of people outside the gate of Dong Hing waiting to be interviewed. Yet ordinary workers were easy to find, but not professional talents. As factories were being set up, Simon realized the urgent need to establish a team of capable professionals and decided to seek talents in inland Chinese cities.

In the early 90s, most professional technicians in China had stable jobs in state-owned enterprises. Even though they lived in a less prosperous city inland and recognized a move to the coast might provide more opportunity, they were hesitant to assume the risk. Simon understood their dilemma and decided to take the initiative. He learned that there were many factories producing towels as well as textile schools in Beijing, Shanghai, Hubei, and Zhejiang provinces, so he went there to seek capable personnel.

"We grabbed our briefcase, took a green train, and set off immediately," Simon said, exhibiting the same drive and resolve some thirty years ago.

Simon liked the metaphor: "running a factory was like fighting a battle with time". He walked very fast, a habit that had developed from his early years of regular exercise. It also showed his personality: quick to act and decisive. When he was young, he could be very bossy and quick tempered.

"He was angry for a good reason, either because of production problems, or low-quality products, or untimely delivery; His anxiety came from his strong desire to make the factory better." Yang said earnestly, "Simon was this kind of boss: he never put on airs; he listened to your suggestions and took them to heart; he also guided you to think." Yang told us that soon after he worked at Dong Hing, Simon talked with him and asked him one question, "Who pays you?"

Startled, he did not dare to answer the question. The answer was sitting right in front of him. Simon saw his confusion and exclaimed, "It is the customer who pays you, not me!"

Yang took charge of pattern design. Good skills were necessary to meet the client's demand, but having the proper mindset was more important. Yang had worked in a state-owned company for many years. To him, this is the first lesson about a market economy: to be market-oriented and customer-focused.

Simon also emphasized the necessity to continually augment their professional skills and to broaden their horizons. In a time when few travelled overseas, Simon took his team to industrial expositions around the world and visited overseas equipment suppliers. He thought his team would benefit from face-to-face communication with industry leaders and from exposure to the most up-to-date technology.

Simon believed that an enterprise should rely on technology, equipment and system to establish brand reputation and seek long-term development. Sustained business growth comes from constant review and a commitment to innovation.

Dong Hing's label design system is a prominent example. There are two steps to make a label: design and production. The process is similar to that of the garment industry. The first step is to design a template. At the very beginning, label design was done manually on an A3 paper, and the production of labels relied on the technician's skill and experience. In the 90s, Dong Hing took the lead in using

❖ *Simon (second from left) and Huang Youliang (second from right) who was recruited from Beijing First Silk Woven Label Factory in 1996. They were visiting Muller in Switzerland and learning professional knowledge about management, label, loom, and pattern design. Huang later became the director of Dong Hing factory.*

❖ *In 2018, Simon reunited with the founding team that promoted the early development of Dong Hing. From left: Sun Dexin, Li Hanwei, Sun Heyi, Sun Jingxian, Simon, Ye Songgen, Yang Yong and Chen Manlin.*

a system for pattern design, which provided a ready vehicle to send the design template to the loom for production.

The software system noticeably sped up the process of designing a label. "What might have taken half a year now required only a few hours," said Yang who was impressed by the high efficiency brought by the adaptability of new technology. Regarding the software of pattern design, there were different suppliers in Hong Kong and overseas. Some loom manufacturer also has its own system, such as Muller in Switzerland.

"I suggested to Simon that we should develop our own software to ensure it was operational on the three types of looms that we used at the factory," recalled Yang. "If I had been working at a state-owned enterprise, I would not have the chance to sidestep my supervisor and speak directly to the boss. Luckily, Simon was very open-minded and asked only how long the development process would take and how likely such an innovation would succeed before giving a green light to go forward."

Simon's decision became crucial to the long-term growth of his company. There was no immediate need for the investment. However, Simon thought otherwise.

"Pattern design is the core of our production process. To globalize our business and production, we needed to ensure that the design process is never disrupted. In other words, we cannot afford to be dependent on one supplier and should anticipate all potential risks. We should take precautions beforehand." Simon added, "And our team had such capability to innovate, with their hands-on experience."

With Simon's support and encouragement, Dong Hing established a development team, consisting of Yang Yong, Roger Chau, Ye Songgen and Sun Jingxian. "We invited the research and development team from Zhejiang University to join us. They knew the basic theory about the module construction while we had direct experience from production."

By melding together theory with practice, Dong Hing was able to develop a system, which was more advanced and stable than other choices in the market. The system was applied to the company's factories worldwide, thus advancing the growth of its global business. "The software system greatly simplified the process. Whereas once it took three years to train a pattern designer, now three weeks would suffice." Yang agreed completely with Simon's proposition to promote development through technology. Such an investment, once seen as a waste of resources, was actually the key to the company's success.

"A traditional business like woven label could also have innovations. The technique system embodied the innovative spirit of the first generation of Dong Hing team, which is very meaningful to preserve and inherit." As the steersman of the company, Simon upheld the daring spirit of innovation and practice, which he regarded as the driving force of a company's development.

With the stable progress of the company's internal system and research and development project, Simon accelerated the implementation of his globalization plan. His target was GAP, the world's popular US apparel brand.

The year 1992 marked the 100th anniversary of the US fashion magazine *Vogue*. The cover of the magazine's special issue featured ten supermodels wearing GAP white shirts and jeans. The outfit, simple, stylish yet classy, shared the popular idea of "making fashion accessible to the general public". GAP became a widely known brand. Four years later, Sharon Stone, a well-known celebrity, wore a dark grey, turtleneck GAP T-shirt on the red carpet of the Academy Awards ceremony. The event proved a catalyst to sales. Soon afterward, the companies in the USA permitted their staff to dress casually to work. The new practice enabled GAP, a leader in casual wear, to develop into the largest apparel retailer in the USA and to expand its reach into Europe, Asia and the rest parts of the world.

"Those who have GAP have everything," Simon astutely observed, assured of the significance of GAP to both the apparel factory and the label supplier in the 1990s.

However, it was one matter to recognize a phenomenon but quite another to take advantage of the circumstance. Many other companies also sought the order from GAP. The whole team in Hong Kong worked very hard to achieve an order from the Far East office of the retail giant. They did receive small ones but nothing meaningful.

"It was very difficult to make a breakthrough from the bottom up," Ronny said. He was the Chief Financial Officer by then. "Simon decided to set an office in Singapore. He meant to build direct connections with GAP's purchasing department in the Asia Pacific without having to go through GAP's Hong Kong office." He added, "I picked Henderson Industrial Park as the office site. It was located in a well-developed industrial park in Singapore."

In addition to GAP, Singapore was the purchasing centre of many other European and US brands. It provided a strategic position in expanding the company's business networks. In 1994, the Group established its first overseas office in Singapore. Simon was cautious and did not want this first step to be overly dramatic.

Two years after the establishment of the office, GAP remained apparently an unattainable goal. The office had not been very useful in advancing the plan. After several failed attempts, Simon started to consider once again the possibility of establishing a direct relationship with the general office of GAP.

On the surface, it might have seemed an insurmountable task, but Simon believed that "nothing was impossible".

Simon had been attending different exhibitions. These visits were one of his many ways to keep up with the latest marketing trends. At one exhibition, Simon heard that AW Printing, a company in Hong Kong, had been making GAP labels. Its factory was in Texas. However, Richard Nip, originally from Hong Kong, managed the factory.

A garment often has five or six labels. That includes a woven label in the collar area, a print label with specific washing instructions and a hangtag, showing price and size. In the early stages, there was a clear division of labour, each producer specializing in one label type. For instance, Dong Hing produced only woven labels while AW Printing, focused on print labels. The division of labour had the effect of eliminating competition. Knowing that there was no threat of added competition, Richard Nip was able to invite Simon to visit his factory. Of course, Simon also has great communication skills.

After their initial meeting in Texas, both agreed that there was great potential for business cooperation between the two sides.

"You should meet Paul Leger." Richard Nip suggested.

Paul Leger, a Jewish American, knew the local culture in the East Coast of America very well. He had built positive relationships with his colleagues at GAP with good business connections in the industry. After leaving GAP, he became an agent and acted successfully as an intermediary between the brand owner and the supplier. Simon immediately recognized Paul's value and decided to visit him in San Francisco as soon as possible.

"You see, Simon knows how and when to seize every opportunity." Ronny recalled, after working with Simon for many years since he joined the company in 1993. "Simon's trip to San Francisco was a critical decision. It was absolutely a milestone which paved the way for our cooperation with GAP." Ronny added with laughter, "Although Simon lost his baggage on the way, his meeting with Paul Leger was very successful."

Simon suffered from a pain in his neck and shoulders around 1995. He had consulted many doctors, but none of them could cure his pain. Simon's friend suggested he learn Ba-gua and Qi-gong, a system of deep breathing exercises to heal his body. Simon was very conscientious and practiced meditation whenever and wherever he could.

While busy with travelling, on his way from Texas to San Francisco, Simon still practiced meditation during the time when he was waiting to board. He found a space beside the boarding gate, put his baggage aside, sat cross-legged, focusing then on his breath, inhaling and exhaling as his master told him. He was concentrating on his meditation, undisturbed by the airport announcement and the passengers going by. He felt the energy current running through his body, his palm warming up and his body relaxing from head to toe. After finishing the whole set of meditation, he opened his eyes gradually, only to find his baggage was gone!

Although he lost his baggage, he found Paul Leger who became a consultant for SML. Paul would be the key to the GAP project. As a master stroke, he decided to set up a factory in San Francisco. AW Printing had, no doubt, not considered such a location.

Simon did not play by conventional wisdom and often came up with what others might regard as fanciful ideas. As a result, in the label industry, he gained the

❖ *During his first visit to San Francisco in 1996, Simon decided to set up a factory there for the GAP project.*

nickname, "Senseless Simon". Even Ronny Ho, who had agreed to hire Paul Leger as the consultant, thought it was risky to set up a factory in San Francisco. However, there weren't many apparel brands on the west coast. GAP was the only major brand with a presence there. Still, 1997 was the year of the Asian financial crisis. Most factories on the alert were operating on an austere budget. The construction of a new factory was a costly endeavour, and Ronny considered the investment too risky.

Simon had his own plan: as a world-renowned casual wear retail brand, GAP offered bulk orders. Further, as GAP's supplier, SML would acquire a quality accreditation. While setting up a factory in San Francisco would increase operation costs, SML could respond in a timely manner to the customer's needs. Moreover, the location near the GAP headquarters would save on delivery time and speed up the approval process when the company offered the pattern design to GAP. But most importantly for Simon, "If GAP placed an order with us, other clients would also come to us. It was worth the risk."

"We have been taught not to place all eggs in one basket. That's wrong! All eggs should be placed in one basket," Simon said.

What Simon meant here with this apparently counterintuitive statement was that one should focus on one matter at a time. And many of his senior colleagues had seen with their own eyes how Simon applied the analogy. Cherry Cheung, who served the company for 30 years said that Simon had "strong reverse thinking ability".

As an example of the laser focus, SML opened up a first-floor office in the same building as the headquarters of GAP in the USA. As a result, everybody working at No. 2 Folsom Street noticed this red SML logo, including the staff of GAP.

In March 1998, SML's factory at San Francisco was completed, only two blocks away from Folsom Street. The 4,000 square-feet space was divided into the production area and office area. Paul Leger suggested launching a classy ceremony for the factory, with artistic style. A band was playing "San Francisco", while a long table was covered with a smooth, creamy white tablecloth. Champagne flutes were neatly arranged in a triangle shape. Buses with SML Logo were waiting for the guests from GAP on Folsom Street. The opening established the image of SML as a fast and efficient supplier, which understood the local culture and prioritized its customers.

Jason Mak was in charge of this factory project. He had worked part-time for Simon's father Suen Shing, helping with finance management before joining Simon's company. As an early member of the staff, Jason served the company for more than 30 years. He took primary charge, during the early years, of building factories overseas. He recalled the first time he met Simon. A young and reticent Simon had accompanied his father to the Kau Kee Roast Meat Restaurant in the Park Hotel at Tsim Sha Tsui. While the father was talking, the son remained quiet. After Jason joined the company, he discovered that, contrary to the initial appearance, Simon was enterprising and decisive. Simon knew how to delegate work and make productive use of his employees. Once he made a goal, he would rely on his team to achieve the objective and would not, unless absolutely necessary, interfere. Jason added, "Setting up a factory in San Francisco was such an important project, but Simon only went there twice, once to meet Paul Leger and the other time, to select the site for the factory."

"Perhaps we had everything in our favour," said Jason. "Things went very well. At the very beginning, we were worried about not finding a good site for our factory, but soon we found one and shortly after completing the tenancy procedure, we began to furnish the factory while recruiting more hands. Whether we achieved a positive start, was dependent on our finding capable operators who were also competent in Chinese. Thus, we did not know, even with an advertisement in a major local Chinese newspaper, whether we would find a sufficient number of qualified employees. We didn't expect then that we would recruit 20 staff who came from Taishan; all were able to speak Cantonese." Jason spoke of this experience with a pride that became noticeable in his voice.

"We hired Paul Sin, a Cantonese immigrant, to be the factory head. Paul used to work in the textile industry in Mainland China and had the required experience and expertise. After that, the whole team was set in place." Things went in line with Simon's expectations. With high efficiency in pattern design, fast delivery and high-quality products, the Group soon attained the globally recognized qualification as the supplier of GAP and began to take bulk orders. SML's success with GAP laid a solid foundation for building a business structure to meet the demand of globalization. The Far East Office of GAP at Hong Kong was shocked at the news that SML obtained authorization directly from the US headquarters. Perhaps nobody believed that such a thing was remotely possible!

In mechanics, a lever is typically a rigid bar that moves around a supporting point called the fulcrum. The bar can rotate about the fulcrum. A lever can increase the input force and produce a strong output force — this is called the law of the lever. The Greek scientist Archimedes deduced this law through geometric methods in his work *On the Equilibrium of Planes*. He once claimed, "Give me a fulcrum and I will move the earth". "Senseless Simon" had actually developed very sensible plans. He had used the factory in San Francisco as a fulcrum and moved the global business. He zoomed in on GAP because he knew once his company became GAP's supplier, it would open up a huge market and more business opportunities. His success perfectly illustrates the law of the lever.

Simon pulled off an impressive victory in this battle, after which SML became a new star in the label industry and found its own footing. With the approval as a global supplier of GAP, Simon began to make plans for establishing global service networks: starting from having sales office in the Far East and gradually expanding the business from Southern China to every corner of the world.

After winning GAP, the US brand, as a customer, Simon turned his attention to the UK market. He picked Keighley, the traditional woven label production site for his factory, and targeted two main UK brands: NEXT and GEORGE.

"We decided on the site in 1998. In 1999, the factory was put into operation," Jason explained.

With the San Francisco factory in fully production, Jason was transferred to the UK where he took charge of setting up the factory. Capitalizing on his experience in the US, Simon cooperated with John Davison, an experienced veteran in local woven industry. SML would provide capital and factory while John Davison would become responsible for daily operation and business growth.

"There're cultural differences. It's better to hire local people because they're more familiar with local culture and know to deal with local customers," Simon reiterated a key principle of his philosophy. As an active thinker, Simon learned from past experience and soon grasped the key to managing overseas sites.

"His success also relied on his precise judgement of the competitive advantage in the industry," Ronny added. "Back then, the agents had the majority of business resources. Connections were keys to more business opportunities. You could not imagine how hard it was for a Chinese person without any connections to expand his business in the West, especially in the very beginning. It was nearly impossible." Ronny explained, "If we sent a local Hong Kong people there as others did, we would most likely not have achieved a favourable outcome. We must trust the local team! Simon also invested a lot of resources into setting up an incentive scheme that upheld 'more pay for more work'. We were not the only one who wanted to go global; the effectiveness of our incentive system constituted a key reason why we

❖ *The UK factory was put into operation in 1999; Simon (left) visited Keighley and John Davison (right).*

were able to surpass other companies in this industry."

Simon was skilled at the art of leveraging resources to his advantage. His strategy was to focus on a key opportunity for growth. Then, he was not satisfied with the accomplishment but continued to build onto that success in order to make SML into a global business powerhouse.

While expanding the global market, Simon remained committed to exploring the Mainland market for further prospects. Once he saw the prosperity of garment industry in East China, Simon sought in 1998 to establish a production site in Shanghai.

Shanghai SML was located at the two-story building on Huaxiang Road, Hongqiao. The semi-circular building was furnished with blue-green glass and a primrose yellow wall. It was housed beneath golden dome. The elegant doors and French windows with the semi-circular overhangs were designed to match the overall style of the building. The three red, eye-catching letters, "SML," were visible even from a distance. No one would expect such a modern, stylish building to hold a label-making factory.

"I'm the designer of the building." Simon said proudly, pointing at his favourite building in the photo. "Even today it still looks stylish."

The advanced design also suggested Simon's management philosophy — specifically, his application of cutting-edge technology based on the principle of efficiency. In keeping with that objective, he transferred experienced staff in Dong Hing to Shanghai, to duplicate the successful business model in south China. Ling Peifang, the first director of Dong Hing who worked in the factory for many years, together with other two: Wu Shunliang and Wang Yuexia, were transferred to Shanghai as the local management.

Wu recalled: "Whoever visited the factory would marvel at the management of our factory. They never expected that a label factory could be so modern." This all owes to Simon. It changed the traditional image of the label industry, blazing a new trail through modern technology and management. Even though the original location is no longer in use, its stylish look and efficient operation remains a source of admiration in the industry."

Dong Hing remained a focal point for the growth of SML. Simon called it the most modern labelling factory with the largest production capacity as well as providing the best living conditions for its staff. His voice was full of pride. "Green lotus leaves stretch as far as boundless sky. Along the clean lake a grand journey begins." His investment had transformed this undeveloped plot into an up-to-date facility linked by a fully paved avenue.

It took Simon ten years to expand his business from southern China to becoming a region leader and finally to a global powerhouse. His success relied on his unique capacity for "reverse thinking"; an ability, in other words, to think outside the box and then, act decisively on his insights. At the threshold of the 21$^{st}$ century, the company, already renamed SML, successfully entered the global arena of the label industry and is marching towards a new century with unstinting dedication.

❖ *The factory in Eastern China, located at Huaxiang Road, Hongqiao, Shanghai, set up in 1998.*

❖ *The factory in Southern China, located at Chang'an, Dongguan, being completed in 2006.*

CHAPTER

# IV

# MAGNIFICATION

/

BOOSTING

# Pioneer before Others, Excel above Others, Change ahead of Others

Zhao Delin of the Song Dynasty wrote in the second volume of *Records of Hou Jing*, "The sun rises and sets like the weaving of a shuttle". "Time flies like a weaver's shuttle" is a perfect analogy to describe the development of SML that began as a company dedicated solely to producing woven label.

Simon knows a lot about yarn and loom.

"There are two sets of yarn, warp and weft, and many different weaving techniques. Depending on the technique, the yarn will produce different textures, which can be seen in the graphics, pattern, density and gloss of the material. The label produced through shuttle loom is soft, but the loom runs at a slow speed; needle loom and broad loom weave faster, but the label they produce is coarse at the edges. It needs to be further processed through ultrasonic cutting and sizing; different production facilities, techniques and processes should be adopted to fulfil the demand of different clients."

"In the 1980s, we used single jacquard air jet loom with 600 rpm; in the 1990s, we improved the efficiency and quality by using double jacquard loom with 1,200 rpm. The colour increased from 8 to 12 and the width to 1 metre. 'Crowd tactics' were never an option, despite the cheap labour cost at that time. The long-term development of an enterprise hinged on advanced equipment and technology."

As an advocate for the modernization of manufacturing who led to a revolution in the label industry, Simon also remained deeply attached to the loom.

The woven label was the core product of SML in the early days. It was one of the two main categories of traditional labels: the other being the print hangtag. The two differed significantly in the type of technique as well as even in the business model employed. Therefore, the woven label and the print hangtag were produced in different facilities. Naturally then, the two types of factories focused exclusively on their own specialized line and did not need to compete against each other.

In the mid 1990s, Simon began to practice Ba-gua. He often said that the guideline for throwing a punch was to be "firm, sharp and resolute". He also followed the same guideline in doing business: "take whatever there is to achieve your aim". To cement the cooperation with GAP, Simon put into practice his philosophy of putting all one's eggs in one basket and set up a factory near the headquarters of GAP in San Francisco. However, such a strategy did manage to provoke AW printing. Here is the background.

Both companies hired Paul Leger as their advisor for GAP business. On the first floor of GAP's headquarter, the two signboards of SML and AW Printing were put up at Paul's agency. In addition, both companies had their own staff working at the agency's office. However, as SML specialized in woven labels while AW Printing in print hangtags, initially, there appeared no conflict of interest. However, SML's business thrived in San Francisco beyond expectations and soon became, at least from the perspective of AW Printing, a competitive threat. They filed charges against SML for using AW Printing's business information without permission.

SML, though a later comer to San Francisco, developed a close relationship with GAP. AW Printing was located in Houston, Texas, at a distinct geographic disadvantage. AW Printing took pre-emptive measures to prevent SML from slicing into its market share, filing charges against both SML's branch company in the US and its headquarters in Hong Kong. The lawsuit was to disrupt the company's daily operations.

"It's not about the outcome but the strategy." Simon said after analyzing the situation. AW Printing knew clearly that SML could not produce print hangtags; however, the purpose of the suit was to nip that possibility in the bud.

Simon considered every detail and made the immediate decision to have lawyers from Hong Kong and the United States to handle the case.

"The most important matter was to keep our company's reputation untarnished. This was the bottom line," said Simon.

AW Printing could not provide concrete evidence. After more than one year, the case ended with no settlement. To some extent, AW Printing achieved its goal. Clients put in orders cautiously — woven labels for SML while print tags for AW Printing.

Nevertheless, this incident served as a timely reminder for Simon. He recognized the importance of diversifying its business to include woven and printed products. Sooner or later, one-stop service would be a must-have.

However, while Simon was focused on expanding and diversifying his business, changing the image of his company as a woven label manufacturer, an internal crisis was brewing at the company.

SML's Hong Kong headquarters at Fu Cheung Centre was a small duplex with less than 20 staff, including Mary, who started the business with Simon.

Ronny's office was on the second floor of the duplex. The first item on his to-do list when he got to the office every day was to check email. As the company's business was globalized, some emails were sent to him at night due to time differences, so he needed to reply as early as possible in the morning. That morning, Ronny looked quite stern after checking his email. He walked straight to Simon's office and closed the door.

He received an anonymous email.

According to the email, several regional heads, led by the one in Singapore, were scheming to start a new business. Some senior managerial personnel at Hong Kong headquarters were also involved. "We were not sure whether the email was reliable but had to take it seriously because the information given was very solid," Ronny recalled. "We did not know the informer's motive, and had no idea at all about how things would turn out. We must prepare for the worst."

After hearing Ronny's report, Simon was very composed.

"Let's go to the airport now-just the two of us."

Facing such an urgent and tense situation, Simon, normally short tempered, remained very calm.

"It just happened that my new passport was right at my office, so I went to the airport with Simon immediately. I didn't even bring any clothes. We got down to business as soon as got off the plane. Only after three days did we have a chance to buy some clothes for a change," Ronny laughed.

Simon did not go to the company after arriving in Singapore. Instead, he went to meet a lawyer and an accountant and told them what happened. Simon followed their professional advice and went, along with his expert team, to the company. Every step was taken cautiously in accordance with legal procedures. When they arrived at the company, the lawyer, on behalf of the company, announced the suspicious misconduct of the local head, suspended him from work, requesting he leave the premises immediately; then they began to search for evidence. In the local head's computer, they found enough evidence to file charges against him. "Ronny was very shocked at what they discovered."

Simon remained calm and reticent, however. This was despite the tropical climate of Singapore whose heat and humidity may sometimes promote an impetuousness and impatience. "We need to catch the snake by its head." He was concerned with identifying the senior manager who was involved and with determining the possible consequences for their misconduct.

Mary in charge of local sales at the Hong Kong headquarters told the suspected senior manager to come for a meeting. He was advised not to bring his cell phone. This was often a requirement at an important meeting. The manager was not suspicious.

Outside the meeting room, the senior manager's phone rang. Cherry, the senior manager, joined the company in 1990 and took charge of administrative work. She knew every staff of the company, including those from the headquarters and the overseas sites. She answered the call and recognized immediately the anxious voice from the other end.

Everything in the email was confirmed.

The whole arrangement, in Ronny's words, was "brilliant."

While Simon remained in Singapore, he asked his wife to hold a meeting in the Hong Kong headquarters and keep the senior manager distracted. He, then, asked the reception to divert calls to Cherry's extension. Simon believed that, if the manager was involved in the scheme as the email indicated, he would be contacted as soon as the conspirators realized their plan failed. Whether Cheung received the call was the key. Everything went as Simon anticipated. And he had already asked his wife to keep an eye on all the documents.

In the 1990s, SML was setting up sales offices around the globe to facilitate the GAP project. Simon did not guard against the possibility that the sales personnel might form a clique to steal the profits, which, by that point, had become sizable. This incident served as a wake-up call for the company's ongoing globalization.

Simon appreciated this senior manager's talent and planned to promote him to a more important position. However, after this incident, he became more cautious in hiring core members. He placed equal stress on integrity and ability, the former even being more important than the latter. Taoist Master Chu Hok Ting, who had known Simon for years, reminded him that he should nip every possible threat in the bud and take preventive measures to guard against potential risks.

Since then, Simon purposefully brought some high-level and middle-level colleagues along on business trips. He said, "Travelling together is a good opportunity to observe your colleagues. At the workplace, what you can see is their work performance, but during a business trip, you can tell their moral integrity by observing how they treat others and act in daily life."

He held regular lectures for staff and invited scholars and experts to share knowledge about enterprise culture and self-integrity; he attended each lecture and shared his thoughts as well. “You cannot unify people’s thoughts, but you can unite them with unified ideas and goals,” Simon said.

Meanwhile, the internal management system was enhanced. A mechanism for supervision as well as a grievance scheme was set up. Simon put a great effort in a variety of measures to raising the level of company performance from developing enterprise culture to improving management system.

“When a business achieves a large scale, how to manage people becomes even more important,” said Simon.

“This is an inevitable path for the growth of an enterprise. Without pain, there will be no motivation for exploration and deep-thinking. An entrepreneur must always be alert to all types of changes, and eventually will benefit greatly from doing so, as the knowledge will be increased, and wisdom improved.” In the face of increased challenges, Simon wrote these sentences in his notebook and, as always, achieved a measure of peacefulness through writing.

Those who can deal with hard times and complicated situations are truly capable. Simon is such an individual.

❖ *In 2001, Simon arranged a lecture for management team on Lao Tzu’s philosophy.*

After having addressed the external and internal problems, he concentrated on the business expansion. Avery Dennison, a US company, approached SML, wishing to establish partnership.

Founded in 1935, Avery Dennison started as a paper-making company with label-printing as its strength. In 1967, it was listed on the New York Stock Exchange. Simon was then just a 10-year-old second-grade student in his primary school, knowing nothing at all about labels. In 1974, when Avery Dennison emerged as one of the Fortune 500 companies in the US, Simon, just 17 years old, had barely resumed his study at Chang'an High School. While he was crossing the finishing line amidst cheers, breaking three records in the school's sports meeting, he never expected he would found a global enterprise and become the rival of Avery Dennison, a top enterprise of the field.

In the beginning, Avery Dennison didn't take Simon's company seriously. "Back in the 1990s, we couldn't conceive of Avery Dennison as a rival. Our business was very different from Avery's." Ronny explained, "Our competitors were the US company RVL and the Taiwan company Dah Mei; both enterprises were engaged with woven label the same as we were."

At the time when Avery Dennison came to seek partnership, it seemed like a perfect opportunity for SML to expand its business by entering the printing industry. The two parties hammered out the details related to cooperation. Avery would be responsible for opening up the USA market while SML would boost production in the USA, assisting Avery in manufacturing. Avery remained responsible for printing, and SML retained its focus on producing woven labels.

The two parties, by 2002, had been cooperating for almost a year. As Simon and his team were about to leave for the USA to review the annual performance of the partnership, his US teammate informed that Avery had just announced its acquisition of RVL. Ronny Ho was shocked, feeling as if he were waking from a dream. While cooperating with Avery, a veteran of the field, Simon realized that his company was inexperienced and relatively short-sighted.

Once the acquisition was announced, the whole Avery team disappeared. At that time, RVL was SML's biggest competitor. The merger, thus, transformed Avery from a partner to a competitor. The partnership between SML and Avery must be terminated.

For a long period of time, SML had only maintained cautious relationships with its immediate competitors who focused on woven label as well, but the jolting experience with Avery, allowed the company to gain a clearer understanding of the industry.

Since the 1990s, economic globalization has increasingly become a factor informing all trade, and as a result, the lable industry, following a widespread trend, has adopted economies of scale in order to remain competitive. Further, major players in the same industry sought to acquire of competitors that would provide a complement in terms of market, technology, and industrial chain and, therefore, enhance the overall strength of the corporation.

This period in US history marked a fifth wave of acquisitions. The label industry had been impacted by the trend. Before being acquired by Avery Dennison, RVL, a world leading company in the label-weaving field, purchased F. G. Montabert, a century-old US label enterprise, in 1999.

"F. G. Montabert, like ASL, was one of the leading woven label manufacturers in the United States," Simon said.

He had visited all its European and US counterparts in the early days of his business and was very familiar with the best enterprises of the industry. Simon was particularly impressed by these two companies. F. G. Montabert and ASL were each large enough to have private jets. Having no successors in his sight, Rollin Sontag finally sold ASL to a listed company that changed the name of the company to ATP, which was then acquired by the US label company Paxar in 1992.

Paxar, also a listed company, was known for its printing and data services. By acquiring a number of companies in the 1990s, Paxar came to hold under its corporate umbrella the major woven label companies in Europe and the US, including Bornemann & Bick (B&B) (German), Ferguson International PLC (UK), and Alkahn (US).

Simon recalled: "B&B, Europe's ASL, was a leader in woven label industry. Until visiting B&B in 1987, I began to realize what automated production was. It seemed that all the outstanding companies had something in common. They considered not only the production process but took into account fashion-related trends in style and colour. Then, based on these factors, these industry leaders designed labels to highlight brand values. It was on these bases that Paxar acquired B&B. The cost was substantial, but the investment was worthwhile."

Simon continued: "Ferguson was a supplier for M&S, a well-known UK brand. After acquiring B&B, Paxar also acquired Ferguson. With these two acquisitions, Paxar had nearly the largest market share in Europe's woven label market. For the US market, the acquisition of ATP (ASL) and Alkahn, two top woven label companies in the US, complemented precisely Paxar's business profile."

"I visited all those peers. I believe those who do well must have something for me to learn." Simon has never changed this idea since he visited ASL in 1985. "It's better to go out and learn from others than to work behind closed doors. If you confine yourself to your own factory, you cannot have a global vision and will never really grow."

At the early stage of his business, Simon's communication with his peers enabled him to have a broad vision of developments within the industry while in later days, such visits further equipped him with brand-new understandings of the strategies required for his company's growth.

During the fifth wave of mergers and acquisitions, the label industry was striving to provide one-stop services; the main players accelerated their pace to venture into horizontal businesses. Acquisition, the most direct and effective strategy was widely adopted. Once setting a target, the listed company was advantaged in having a ready of funding to finance the expansion.

"Paxar is model for the strategy," Simon noted. "It leveraged its strength as a financing platform, acquiring its most capable peer companies. That's why it became a success." Simon had once visited Paxar's New York headquarters and spoke highly of Arthur Hershaft, the President of Paxar.

SML faced the same wave of mergers and acquisitions as well as a series of internal crises at the beginning of the 21$^{st}$ century. These events inspired Simon to reconsider the future development of his enterprise. His focus subtly shifted from an emphasis on production and sales to developing a coherent strategy for expansion. He eagerly searched for acquisition targets to enhance the comprehensive strength of his company.

From then on, SML's rivals were no longer restricted to woven label manufacturers. Simon began to keep an eye on counterparts that were known for their printing and data services, such as Avery, Paxar, and Checkpoint. Thinking of these rivals, all US listed companies, Simon asked himself: "How can we catch up?"

Acquisition was nothing new to Simon. Not long after he started his business, he acquired a printing label factory. However, unlike the "1+1=2" idea in the beginning, Simon's acquisition targets became those who could help his company to achieve the "1+1=3" synergy effect. After Avery acquired RVL, Simon realized his top priority was to acquire a US printing company to strengthen the Group's marketing

❖ *Simon (middle) visited Paxar's headquarters in New York with Arthur Hershaft (second from right), President of Paxar.*

capability in the US. Although the business with GAP had developed very well, the retailer was not located primarily on the west coast, which was not a hub for apparel accessories. It became incumbent for SML to establish a foothold on the east coast, thereby expanding its share of local market as well as deepening the footprint in the US. Simon pinpointed two keywords to describe his strategy for acquisition: "printing" and the "East Coast".

While he remained driven to deploy efficiency as a force enabling SML to catch up with some top enterprises of the field, Simon remained quite composed, believing: "Subjective wishes should obey objective reality!" Simon had a blueprint for success. First, he hired Fabio, an expert in the US label industry, as SML's advisor. As such, Fabio became responsible for identifying and recommending suitable targets for acquisition. It was vital to locate a local expert.

Fabio was a plump American and frequently wore a smile. He had a well-established reputation at being a skilled negotiator while remaining easy to get along with. He had a broad network in the industry. With Simon's clear objective and direction, Fabio was able to locate with expediency an acquisition target: Kalpak, a

US company boasting woven and print label technologies. Although based in New Jersey, Kalpak had an office in Hong Kong, where SML was headquartered, this proximity represented an advantage.

A simple yet grand dinner was held at a small estate in New Jersey, to celebrate SML's successful acquisition of Kalpak. This was the first acquisition in the history of SML, and Simon saw the event as a milestone. He decided to give a speech in English at the ceremony.

"The acquisition will significantly improve SML's production and operation capacities in Central America, South America and the Caribbean, and raise the status of SML in the label industry. The strength of SML will also be enhanced in the US, particularly on the east coast," said Simon.

The acquisition of a US enterprise by a Chinese company drew a great deal of attention. Textile World, a US media, cited the above two sentences from Simon's speech in its report on this event.

As usual, a badge with the SML logo was pinned on Simon's suit. After his speech, he walked to the table and greeted his guests. Ever since began to practice martial arts, he had been keeping his hair a little bit long. He actually looked like a typical person from the east. At the end of the party, he raised his cup and said "cheers" to the guests. Meanwhile, deep in his heart, he made a toast to himself: "A small step for him, a big step for SML." Perhaps only Simon as its founder understood the true significance of his speech in front of almost a hundred guests.

Talking about the difficulties encountered at the beginning of SML's overseas expansion, Simon said: "Although we were very successful in San Francisco, we met lots of difficulties in New York, as we failed to attract qualified local employees. Some potential employees came to us but left immediately as they learned that we were a Hong Kong company. For sure, they preferred to work in a well-known US firm. No matter how competitive the remuneration we offered, they just didn't trust us. That was the reality." As usual, Simon saw the matter practically. "The only way to attract capable people was to make our company sufficiently powerful. Instead of trying to change how others thought about us, we'd better improve ourselves in the first place."

"The market will soon realize that our company is a very good one. Business competition is all about strength," said Simon confidently.

He was right. Thanks to the successful acquisition of Kalpak, SML finally had its business in the east coast and earned a name in the US market. Gradually, there

❖ *Simon (middle), the then US general manager Geoffrey (left) and SML's advisor Fabio (right) on their way to visit potential companies for acquisition. New York, 2003.*

❖ *Simon celebrated the acquisition of the US company Kalpak by proposing a toast among the staff after his speech in 2004.*

were talent individuals from the surrounding community, willing to work in SML. And Simon's speech, at the ceremony, impressed his staff a lot.

Gary, in charge of the then Kalpak's financial and IT departments, is now the Senior Vice President in America, mentioned Simon in his exclusive interview with America's *CEO* magazine in 2019. He spoke highly of Simon as an entrepreneur who had turned a Hong Kong company with only a dozen people into a global enterprise with footprints in more than 30 countries. Gary described Simon as a visionary and innovative entrepreneur, breaking through in the face of immense challenges.

After the acquisition of Kalpak, Simon immediately adjusted the internal departmental structure in response to the diversification of his business. He also appointed PWC, a Big Four accounting firm, to conduct the audit of his company. Simon was fully aware that the recognition of SML's financial figures by professional institutions was significant for running a global business. "Although our cost increased, we successfully avoided mistakes that may occur if we relied solely on our internal financial management. This approach also helped to shape our company as a truly professional one." Simon recognized the long-term advantages to building a sound financial structure.

After establishing offices in the United States both on the west and the east coasts, Simon turned to Europe where he planned acquisitions that would further develop SML's print sector, expanding the business to the global market. Before Asian brands emerged, such as Japan's Uniqlo, the major retail brands for apparel were situated in Europe and the US. Based on his previous experience setting up woven label production overseas, Simon decided to concentrate on the UK, which he determined was the hub of the European market.

"As long as there is a market, there is an opportunity."

The next year, SML acquired STR Gresham, a UK digital printing enterprise.

It was summer, July 2005, the signing ceremony of SML's acquisition of STR Gresham started in Corby, a small town not far from London. The tables in the conference room were arranged in a circle. The creamy-white tablecloth was in positive harmony with the flower arrangement — pink carnation, purple lavender, white hydrangea and lily, presenting a fresh and graceful atmosphere. Simon wore a black suit and a light pink tie with white polka dots — an interesting combination of colours. He looked dignified, also a little bit playful. As he watched the other party signing the document, his eyes full of joyful lights. When got up to shake hands with the representative, he looked at the camera, happy and relaxed, with the calm confidence of a global entrepreneur.

*The British Printing Weekly* wrote that this acquisition enabled SML to be one of the largest label suppliers in Europe. SML UK would operate as a service centre in Europe, providing data management services, variable labels and other products for European fashion brands.

Simon kept the time-worn pen in a drawer of his bedside cabinet. This pen was used to sign the contract — a milestone in the history of SML. The rusty pen shaft showed the passage of time.

Those who hope to achieve greatness must comply with the trend and act accordingly. Simon was fully aware of the trend of economic globalization and the customers' collective desire for one-stop service. He also understood clearly the direction where his competitors were heading. In the following two years, he continued to act the trend towards one-stop service and acquisition, increasing the market share in the US and the UK. He expanded his business in Europe, where he established offices in Germany, Italy and Morocco. In this way, he was able to absorb

❖ *Simon (right) at the signing ceremony of acquiring the British company STR Gresham in 2005*

❖ *The pen Simon used at the signing ceremony*

❖ *Simon (middle) acquired Bell Manufacturing, a US Company.*

diverse talent, technology, reaching thus a broad customer base. As an outgrowth of the expansion, SML was able to strike an effective balance between the woven and print business. The global network was taking shape. As a result, SML was achieving a market evenly weighted throughout the world.

After a series of acquisitions, Simon prioritized the endeavour to optimize of the EPR system and to establish an E-platform. He was aware that whether the success of his acquisition strategy in effecting synergy relied entirely on data processing and managing capabilities. Established as a woven label manufacturer, the company had not, however, developed sufficient capability to manage printing-related data.

Ronny unfolded the story: "We have been working on the ERP system for a long time and hired a software company in 1994 to help us develop the two systems — an order management and accounting system. Both were actually basic ERP applications. In the beginning, we needed to connect Dong Ying and Dong Hing via the ERP system. The process is complex, and we also hired a professional consultation team to take charge. Later, when Roger Chau joined our company, we recruited some IT professionals and built an IT team. The ERP system was then widely used in our company. However, the system was mainly intended for woven business, which was not an efficient tool to handle printing business. The intrinsic difference between these two parts contributed to the system inefficiency."

In addition to the difference in production equipment and manufacturing techniques, another major difference between woven and print products pertained to the customer base. For the print business, it mainly served brands. As soon as market-related data was obtained, production could begin. For woven business, however, orders were placed by garment manufacturers, so the order needed to serve the customer's specifications. The biggest difference between the two was that print products covered much more data than woven production. Thanks to SML's acquisition endeavours, orders fooded in the system. As a result, SML greatly extended its reach in the global market. Against this backdrop, it was urgent to optimize the whole system. The key was to properly integrate orders, control panels, and operational standards into the system.

Previously, Simon had set up the ERP system to conform to an automatic or computerized management system. However, as SML had grown significantly as a global entity, his newly installed system was set up to be benchmarked against global standards. This upgrade of management system was in line with Simon's plans to enhance the data processing capability all around the global sites and facilitate its transformation from woven to printing.

Simon precisely grasped the key point.

Simon has a firm belief in the power of technology to enhance efficiency; thus he attached great importance to the optimization project, forming a team including two IT experts, who were from Target. Target was a large retail enterprise in the United States second only to Walmart. Simon wanted to leverage the two professionals' rich experience and their data processing ability in the retail industry to optimize SML's ERP system. He abandoned the old model. He also invited Infosys to take care of the system design. Infosys, listed on the New York Stock Exchange, was a top-3 software enterprise in India. It specialized in software development and system integration and provided solutions for enterprises.

"The combination of America's vision and India's software was superb." Ronny, a trusted employee who worked with Simon for 14 years by then, immediately grasped his conception. Simon had great hopes for this highly international and professional team. However, the results were not always up to his expectation.

A label may appear to be a small and simple item to produce, but the makeup of the industry is actually quite complex. The production of a label may involve a lot of variables and data, such as size, colour, and price. Further, the size of the transaction involved in each order can be relatively small: what is called a "fragmentary order" in industry parlance. The key is firstly to make clear the demands of customer and the production requirements of each factory, and then make the system reflect them all. Communication is, of course, a core element of a system. The individual stakeholders in the process must understand and be responsive to each other's needs in order to produce varied products that are, nevertheless, up to the same standards.

"Initially, communication was not smooth. This may be due to language barriers", Ronny reflected. "Our foreign management staff couldn't communicate very well with the factory workers. In turn, the requirements at the frontline were not accurately reflected in the ERP system."

As soon as the system was launched, SML received a number of complaints. The more alterations technicians made, the more problems it incurred. Ronny said: "Many settings were actually interrelated. The change of one setting would affect many others and may even alter the whole system. The more we modified it, the more complicated the problem became." Instead of bringing efficiency and benefits, the optimization project became a large obstacle. This was an unanticipated situation.

For the refinement of the ERP system, which failed to achieve the desired effect, Simon admitted that he had to bear the major responsibility. "My hope was that we could set up a system integrating elites and professionals from Mainland China, Hong Kong and overseas, so that their strengths could be combined. However, I overlooked a very important factor." Simon paused and raised his voice, "Culture!"

"Cultural difference produces differences in thinking patterns. Cognitive gaps and language barriers could easily lead to misunderstandings. The whole project was led by a foreigner, who not only failed to reconcile all the differences and create a harmonious atmosphere but made the project simply difficult to carry on. Indeed, international experience was good, but without adjustments to local realities, that experience does not necessarily yield positive results."

The ERP system optimization project, costing almost 50 million HKD, had to be called a halt. In the context of the 2008 financial crisis, the failure of the system was especially significant. For a label manufacturer, one order may produce a profit of less than 10 Hong Kong dollars. As a result, the capital chain of SML was disturbed. "It was a great setback and a serious lesson for us." Simon never avoided the subject. He mentioned the project several times, noting that it was one of the most profound lessons he had ever learned in promoting technology and modern management. "The cost was too high," remarked Simon.

In the face of such setbacks, Simon believed that the most important matter was to avoid negative emotions. "Each setback can give us a chance to reflect on ourselves. We can ask ourselves some questions. For example, which factors did we ignore? Was the failure caused by design or execution? We need to analyze these questions carefully and find out the fundamental reasons behind. Only by doing so can we learn from our errors and take more effective action in the future." Simon was, as always, positive and sharp-eyed. "The loss of money was a profound lesson we learned. We suffered from pain, so will be cautious next time. There are no fixed rules in managing a business. We all need to learn from mistakes and failures."

K.C. Lau, who joined SML in 2007, was initially the Senior Vice President for data service. Now he is the CEO of the company.

Lau recalled: "I met Simon for the first time at Dong Hing's Spring Dinner. I was not a staff member of Dong Hing back then. When the dinner finished and everyone was going downstairs to leave, I was introduced to Simon, who paused to look at me. After a noticeable period, he shook my hand and used the word "upright" to describe me. I will never forget that moment."

In judging a person, Simon has his own way. The first project assigned to the "upright" Lau upon his arrival at SML was the optimization of the ERP system. This was an important project indeed but had been somehow set aside for quite a while.

"As soon as took office, I wrote an evaluation report on the performance of all our branch offices around the world. I found that the data service system of our company was fragmentary. For example, one system for UK, another totally different one for Hong Kong, but there was no system for the US."

Lau had an IT background and was familiar with the printing industry. "To be honest, the talents from the companies we acquired were all excellent," Lau said, "but they didn't know how to make concerted effort to achieve the same goal. In my report, I mainly suggested restructuring the team, and clarified labour division and cooperation among the different departments."

After reading the report, Simon gave Lau a green light.

In three months, Lau had carried out his plans. Unsurprisingly, the integration of data services brought about the much-desired synergy effect. Without hesitation, Simon assigned Lau as the team leader responsible for the optimization of the ERP system.

Simon's decision proved to be correct.

The first thing Lau did after taking charge of the project was to figure out why the initial launch of the system was a failure. He replaced Infosys with a local supplier and worked out a detailed timetable and to-do list. He also trained his team members to make sure that everyone understood the functions of the new system.

"This was a big challenge for me," said Lau. "Before that, all I did was to implement tasks assigned by my superiors. But this was the first time I took charge of a major project." Lau confessed that he was a very ambitious and competitive person. After he was entrusted with this important project, he was determined to be a good team leader.

"During my stay in Dong Hing over the two months or so, anxiety was with me every day. I communicated with my colleagues to try out the system again and again. Dong Hing was the primary site. If we could do well here, it would be easy in other plants."

As a leader, Simon showed the capacity to discern his staff's potential. In early 2008, the new ERP system was officially launched, and the old system was abandoned. Shortly after that, Simon met Lau and gave him a pat on the shoulder, saying: "The new system is really a success! The sales increased dramatically."

"Hearing his words, I was so relieved." In his office, Lau recalled this period with a smile.

After the successful launch of the ERP system, Simon asked the team to develop yet another one — the E-platform. His standard was clear — pioneer before others, excel above others, and change ahead of others. Only by doing so could his company stand out among the competitors.

From the beginning of his career in the label business, Simon sought these same advantages. A colleague of Dong Hing, responsible for pattern design, recalled, "Simon usually brought some well-woven labels for us to learn. These labels were produced by other companies. He requested us firstly 'can do', then 'do better', eventually 'do differently'. That's how we gradually became the best."

Guided by this philosophy and aspirations, SML launched an online order system for customers. With several new and innovative functions, this system, upon its release, became extraordinarily popular and the internal efficiency was also improved.

Using care label as an example, in the past, the routine would involve pre-print, product development and customer service departments, to handle the order, including design and seeking approval. The whole process took seven to ten days.

In order to address the serious issues with time, a new function was devised for the online system. Once relevant parametres and requirements were set in the computer, the system automatically showed a sample image of the product. A click of the mouse on the customer part matched the expectations. The whole process took only one day. What's more, these customer-approved samples could be shared with different production sites across the world. Gone were the days when a sample was sent from department to department. This streamlining not sped up the approval procedure, but the increase in efficiency allowed SML to expand its production capacity.

SML's new factories benefited a great deal from system optimization as well. "It's like cooking. We no longer need to teach the new sites as the online system contains clear guidelines. This became our recipe for success." Lau used a plain metaphor to describe the new system. "When the new sites were set up in Southeast Asia later, we don't need much manpower to process data. The overall operating efficiency became greatly improved."

"When SML started print label business, our customers were understandably suspicious. While they were well aware of SML's excellence in woven label, they would wonder if we were up to the task." Ronny confessed. However, as the E-Platform and the ERP system were applied to all SML's global branch offices and factories, data-powered businesses, such as printing, became fully supported. Very soon, the high quality of production dispelled the suspicions. Over the next decade, SML's printing business expanded exponentially.

After a substantial period of acquisition and optimization, Simon achieved his aspiration. Starting as a woven label manufacturer, his company has become capable presently of producing efficiently high-quality print and woven label. It also boasts one-stop services for a global market.

During the close of the 20$^{th}$ century, the economy of East Asia underwent a rapid change. Hong Kong, Singapore, Taiwan, and South Korea began to develop rapidly exported-oriented industries, relying heavily on labour-intensive modes of production. The strategy was highly successful. Their economies soared, and they were nicknamed the "Four Tigers of Asia". Similar to other enterprises in this region, Hong Kong's garment industry began to flourish, and the label industry was led, likewise, by its "four tigers": Joint Tag, Brilliant, Shore to Shore, and Dong Ying (later known as SML).

Patrick Lau, the owner of Joint Tag, had originally been an accountant and had a numbers-driven approach when it came to running a business. He did seek synergy or what was called in company parlance, a "1+1=3" effect. However, his focus remained on the immediate bottom line, not on the growth of the company in the long term. For him, the profit must be right in his pocket, not anywhere else.

Brilliant, another "tiger" of the label industry, sought primarily to serve US customers, and the strategy was highly successful, leading to the label supplier to become a giant. However, unlike SML, Brilliant chose not to invest in an advanced system and remained as an OEM supplier. Unlike Brilliant which operated more impersonally, John Lau, the boss of Shore to Shore, sought to manage each detail of his business. He believed in the idea of "no effort, no profit". He was forever moving from one location to another. In the eyes of many professionals, he was very diligent. John was always armed with several mobile phones. He put his family members in all the main positions of his company. For a while, Shore to Shore was on an equal footing with Dong Ying in terms of sales volume.

All these companies were primary players and rivals in the market. Each had a leader with a distinct managing style. However, with the notable exception of Dong Ying, the companies all saw their market share recede. SML, as it became later known, remained headquartered in Hong Kong but managed to venture into the broader Asian sphere before venturing out onto the global market. Simon successfully transformed SML from a regional label manufacturer targeting local market to becoming a transnational corporation providing one-stop services.

"Pioneer before others, excel above others, change ahead of other," Simon says, summing up his philosophy. When services in this industry become good enough after sufficient competition, he must find a way untrodden by others so as to add something new to the industry. "Change" is the key to Simon's business philosophy.

"In the beginning, 99.8% of our company's revenue came from the woven label businesses. Had we not expanded our business, we would have never enjoyed the status today," said Ronny, who admired Simon's managing style very much. "Simon is extremely sensitive to the changes of the industry and pushes the company to change with them. He is also aware of the necessity for adjustment in the process of adaptation. There were a few false steps but quickly returns the right track."

"Actually, we paid a high price in our transition from woven to printing. Seeing the failure of our competitors to make the transition, we became extremely cautious, in particular, to keep on good terms with CDS's technicians. Otherwise," Simon joked, "the million-dollar machine may lie there and rust."

When Simon just began the print label business, he bought the silk printing machine from CDS, an Italian brand. Since all the fittings of the machine were exclusively possessed by CDS, he had no other choice but to rely on CDS's technicians if any problems occurred.

Often, change implies difficulty and sometimes may lead to failures such as the initial launch of the ERP system. However, if the entrepreneur remains flexible and ready to learn from a mistake, he may drive a business to make breakthroughs in the global marketplace, which can only be achieved by "adaptability".

This adaptability not only informed Simon's managing style, but also his strategy in acquiring other companies. As a private firm, SML did not possess the advantage of being able to integrate different financial channels. However, to win more with less was a test of Simon's business wisdom.

If we say that Avery Dennison's acquisition of its rivals was like eating "big" fish, then SML's acquisition was more about eating the "right" fish. That is to say, the scale of the company to be acquired was not the priority to be considered. Rather, Simon was concerned about whether they could help overcome SML's shortcomings and complement SML's existing businesses. After SML acquired several companies of the field, some other companies came to SML, hoping to be acquired. Of course, Simon had his own evaluations of these choices.

"Many companies came to us. Trimco was a good company, but it didn't serve objectives of SML to provide one-stop service, so we didn't go ahead." In 2012, the Partners Group, a Swiss private equity, acquired most of Trimco's equity shares. Three years later, with the support of Partners Group, Trimoco acquired A-Tex. "Although Trimco's performance turned out satisfactory later on, I don't believe the acquisition represented a missed opportunity, for one needs to think about long-term effect, not momentary profit."

As SML became strong enough through strategic acquisitions, Avery Dennison became more aggressive. With adequate capitals and ample experience, Avery Dennison acquired RVL in 2002 and Rinke Etiketten, a German label-weaving company, in 2004. Its acquisition of Paxar in 2007, also a listed company, by using 1.34 billion dollars, created a sensation in the industry. The next year, it proceeded to acquire Dah Mei, a woven label company based in Taiwan, hoping to become a superpower of the field. Hsu, the owner of Dah Mei, once had a casual talk with Simon at a textile exhibition many years ago, who immediately recognized Simon's vision. Hsu told his successor, his son James who held a doctorate in chemistry, that

❖ *Simon (middle) visited A-Tex, a label company in Denmark, with Jan Jakobsen (left), CEO of A-Tex, and Bent Povlsen (right), founder of A-Tex.*

❖ *Simon at A-Tex's demo room*

Simon would be his biggest rival in the future. However, Hsu didn't realize that before Dah Mei became powerful enough to compete with SML, Avery Dennison would have already acquired the company.

The first decade of the 21st century was drawing to an end, and the period, when companies, despite their size, could compete against one another, had also come to a close. After a series of acquisition and purchase endeavours, label companies were totally reshuffled. Only the fittest and most powerful companies survived — only those who could provide one-stop services to the global market. Using huge amount of resources, Avery Dennison reached its objectives in terms of acquisitions. It became No. 1 in the field.

Nevertheless, from his beginning as an athlete, Simon has striven steadily and powerfully towards success. And SML has gradually surpassed other competitors, most notably, Checkpoint, a listed company in the US, jumping to the second place in the field, only one step away from Avery Dennison. Undoubtedly, SML has been a blossoming force in the label industry.

SML

CHAPTER

# V

# TRANSFORMATION

/

## MAKING CHANGES

# Emphasize the Reality and Recognize the Changes

Simon's friends introduced him as the "king of label". In case the significance of the title is not immediately apparent, his friends would also point at the label inside their suits and say: "Well, this is exactly what Simon produces. His factories are all around the world!"

The statement often provokes confusion. Why does label production require a global network? That question comes up frequently. It is not readily apparent that a company producing a small item needs to be a very large concern. Rather than attempt to clear up the matter, Simon remains silent. His reserve is a long-standing trait. In his old days in primary school, Simon had been silent when punished for tardiness despite his sound reason for lateness: that he was busy helping his family with the farm work.

Instead of attempting to dispel the false belief, at the virtual store in the technology innovation centre in SML, Simon spoke earnestly about the future of the label industry: "The main function of label will be the carrier of consumer technology."

Eyes beaming, he randomly picked up a shirt on the shelf, held it high and drew the attention of his audience to the label. Looking at the item in the light, one can easily see a small silver-coloured dot right in the middle. "This is a RFID (Radio Frequency Identification) chip," explained Simon. "Once you have it, each commodity has a unique code. This is very important for the retail industry in logistic management. It can help to track and trace in real-time." He put the shirt back to the shelf and added, "This is just a very basic application of RFID."

"This innovation will change the business operation, as RFID labels can be used for stock management, replenishment, self-checkout and managing the entire supply chain by digitalized platform." He continued: "Nowadays, a great proportion of orders are made online. BOPIS (Buy Online, Pickup in Store) or BORIS (Buy Online, Return in Store) is quite popular among consumers. With RFID label, all inventory data can be collected and visualized in a central platform, which improves the operating efficiency significantly. This is called Omni Channel. It means that online and offline channels are fully integrated."

Recently, Simon frequently meets investors. "As listened and talked a lot, now I can elaborate this technology application well." Simon smiled. Obviously, he is a fast learner.

"Ultimately, the combination of Big Data and Internet of Things (IoT) will be the trend. Artificial intelligence (AI) can help us analyze the data in the system. On that basis, IoT can simulate and predict consumer behaviours. This innovation will significantly challenge traditional mode of production and consumption and even pose a challenge to the retail industry as a whole. The consumer technology will be a driving force for business growth in the future," Simon raised his voice.

"We have also explored other aspects related to consumer technology," added Simon. "Like robots having already been successfully developed in the innovation centre in Shenzhen. With RFID labels and solutions, these robots can do a stock check automatically."

Had it not been for Simon's live demonstration of these applications, it would be hard to imagine that he was describing a label manufacturer. A label is small indeed, yet it can be very powerful.

"Yes, we are in transition," admitted Simon, "We are striving to become a consumer technology solution provider." He quickly responded noticing the signs of disbelief registered on the visitors' faces.

"The era changed, so did the environment. Those who can take the lead in the market must be, in the first place, a leader in technology. We must emphasize the reality and recognize the changes." Simon paused, "For this point, I once made a mistake".

Determined to practice martial arts, Simon has for years been an early riser for morning exercises. On an ordinary Monday, Simon rose up, swam, and practiced quick walking. It was 24 June 2019. He then sat down at the table about to have his breakfast, which was simple — a boiled egg, a cup of milk, and some sweet potatoes. He had small but frequent meals every day. Even on these daily trivialities, he was disciplined in practice. He unfolded a newspaper, and became attracted immediately by a headline, "The Biggest Mistake Made by Bill Gates". According to this report, Bill Gates confessed that a greatest error he had ever committed was not to urge Microsoft to develop a standard, non-Apple system. Google recognized the opportunity and filled the gap, developing Android. It became the king in the non-Apple smartphone market.

Simon sank into meditation. A past experience hidden deep in his heart now resurfaced.

As early as 2004, SML's competitor, began its foray into RFID, Simon had been dedicated to rapidly expanding to the print label business through a number of acquisitions, aiming to provide one-stop services. In searching for target companies, he became drawn to Alltag, a company that had dual tags which boasted both RFID and EAS. Alltag was not apparently on Simon's target list of acquisition. Nevertheless, he remained interested in Alltag and in 2006, after successfully purchasing some UK and US printing corporations, he went to meet the CEO of Alltag.

The CEO, a tall European gentleman, was a great dresser and, asked for a purchasing price twice the company's annual revenue. "Money was not a problem. We did have the capability to purchase it", Ronny recalled. He joined the visit with Simon. However, the primary concern was whether, for SML, the purchase would have synergistic effect. "We thought that the timing was not correct. For one matter, while the application of RFID to retail industry was a novelty, its prospect was unclear. Further, we didn't have RFID labels and any clients at that time."

Both sides have not met since the initial discussion.

Five years later, as Simon ventured into the field of new retailing technology, he thought of this experience, considering whether he had overthought the matter, and missed a golden opportunity. EAS and RFID technology formed a perfect combination.

Later, a company in Europe purchased this patent from Alltag. The price remains unknown.

"In the world of technology, winner takes all. It is very difficult to make a right decision. However, a right decision can change all," Simon said emotionally, his eyes still fixed on a computer screen. For him, buy the right thing is the only right thing to do. That the value surpasses all.

Objectively speaking, SML achieved great success in its acquisition of two US companies in 2012 and 2013 respectively. In 2012, it purchased CGP, a company in North Carolina that primarily focused on RFID label business. In 2013, SML purchased the Texas-based Xterprise, a company supplying RFID logistics management software. CGP and Xterprise together augmented SML's capacities in RFID production, market network, and software solution. These three factors are keys to growing RFID business. Simon grasped the core aspects and had very clear targets in mind.

While the acquisition project in the US went smoothly, the company proceeded,

❖ *Simon (middle) acquired CGP and developed RFID business in 2012.*

then, to establish a technology innovation centre in the UK to develop corresponding business capabilities in Europe, followed by building RFID production capacity in the Asia-Pacific region. The expansion on business as well as global footprint undoubtedly provided a wonderful combination.

"We need to know how to balance the development of various regions. Relying on a single site absolutely won't work in the long run. Distant water cannot put out a nearby fire. Different strategies are needed for different sites. One must plan for all contingencies, and 'when possible'", Simon added, "I preferred taking the path of least resistance."

"The UK has talents, and a solid business base and, more importantly, the market. Brands in the UK are the most open to RFID application." As a result, he decided to set the European base in the UK.

Having practiced martial arts for many years, Simon is well versed in the fist technique, a practice grounded in a series of conventional moves. Likewise, he was aware of the importance of finding the method suitable for a particular context. "In the UK, it was most feasible to set up an innovation centre, applying CGP technology to leverage local talent to capture business."

After the establishment of the innovation technology centre in the UK, the company has successively obtained approvals from several well-known UK retailers, providing RFID tags and solutions. In turn, the business on the continent gradually grew. A good beginning suggests the smooth growth of business. Being "far-sighted", Simon brought about the stability and prosperity of his company.

In the Asia-Pacific region, Simon focused on developing RFID capacities.

Since the scale of CGP was limited, the focus should be placed on replicating production experience regionally. Simon remarked, "The market is in Europe and America while the production should be based in the Asia-Pacific region. Both market positioning and labour division must be clearly demarcated. Dong Hing has the best production capability. Therefore, other production sites could look at Dong Hing as their role model if it succeeds."

Having experience as a teacher, Simon has his own teaching model for operating a company. He believes that for a multinational company with more than 20 production sites around the world, a one-time full-scale roll-out requires a lot of resources and carries high risks. It is more effective to find a key factory to perform a pilot and after the initial success, safely replicated the practice worldwide.

K.C. Lau observed candidly developing RFID production in Dong Hing as a "start from scratch" process.

"After acquiring CGP, we received approval for RFID projects. At that time, without sufficient ability to produce our own chips, we had to find a supplier to ensure that the raw materials were available on time and in a suitable quantity. Additionally, we needed to get equipment in place, such as converting and encoding machines, all of which were sophisticated and expensive. If the desired results were to be achieved, a great deal of preparation was needed, yet the team was committed. Initially, everything went very smoothly."

But technology innovation often contains an element of unpredictability.

Not long after it was installed, the converting equipment suddenly broke down. No one at the factory had used the machinery before, finding the supplier directly to check and fix seems the only safe play. While it was Christmas, and the professional technicians in Germany could not come to the site until after the holiday: mid-January of the following year. The factory had no choice but to wait. The situation was made even worse when an unexpected notice from the customer came, suspending all the orders of RFID products. Lau recalled, "Our colleagues had been preparing for the new project with great enthusiasm. It's like a bucket of cold water poured on the top of their heads, before they could recover from the first shock."

The whole team was very frustrated. "It's like you are rushing forward, but the referee suddenly blows the whistle, saying that the game has been cancelled. Felt very bad."

When the first battle was set back, Simon only said, "Don't lose heart when you are dejected; don't be conceited when you are successful." He reminded the team not to lose confidence and at the same time, to reflect seriously on the whole process. Simon remarked, "The client's decision couldn't be controlled, but the risk could be contained." After receiving approval for the RFID program, the team had been too optimistic, neglecting the actual performance of the client and their hesitancy in applying RFID. Simon had hit the nail on the head.

Fortunately, the company successfully obtained another RFID approval from a well-known sports brand in 2013. The RFID production line in Dong Hing went into operation. In the same year, the acquisition of Xterprise was completed. The market position of the company has since grown more well-defined. With RFID labels, the company could provide a one-stop technology solution for inventory management and theft prevention.

At the dinner celebrating the acquisition, Dean Frew, the representative from Xterprise, presented Simon with an oil painting by a Texas painter: cowboys on horses galloping across the wilderness. Simon returned with his own gift: a woven painting made by a Dong Hing loom, showing a flock of white cranes flying over the lake and mountains. The two paintings, in tandem, illustrate the company's development philosophy. On the one hand, the company demonstrates a frontier spirit, driving business to grow through a forward-looking application of technology; on the other hand, the patience to consolidate its foundation is a must on the road towards seeing whether RFID business can grow profitable. The support from the woven and printing sides of busines form crucial pillars underpinning the company's success.

Simon's business philosophy is deeply rooted in the traditional culture, which emphasizes harmony and balance in all aspects of life. As Simon notes, "Technology and traditional parts are interwoven for the corporate growth. Without the presence of both elements, there will be an imbalance. That's why we started to expand our traditional business in Southeast Asian countries long ago."

When discussing how to grow business in Southeast Asian region with the management team, he often made reference to his experience in the 1980s when building the factory in Dongguan. He pioneered the industry by applying electronic looms, the latest innovation at the time in textile industry. Simon added, "I don't mean to show how smart I was at that time, but to remind the team that technology is also indispensable for a traditional industry such as labelling. Though labour resources in these countries were plentiful, there was no need to adopt a 'huge-crowd' strategy. Automatic production was essential to efficiency." Relying on technology remains a key to achieve of sustainable success.

A book was half-spread on Simon's desk. His secretary had just brought over the latest Bangladesh investment guide published by Bangladesh Consulate General Hong Kong, which was divided into two parts: one introduced the local investment environment; and the other featured interviews with the current foreign investors in the country. The page on SML was specially folded, revealing only the SML Logo, the white lettering enclosed in a red background.

"We're probably one of the earliest companies that took advantage of the opportunities presented by the Belt and Road Initiative." Simon said.

At the beginning of the 21st century, Simon has set up factories in several Southeast Asian countries such as India, Vietnam, the Philippines, Bangladesh, Sri Lanka, Indonesia, and Cambodia. When National Development and Reform Commission released *The Vision and Actions of the 21st Century Maritime Silk Road* in 2015 to specify the areas of coverage for the "Belt and Road" construction, the company had almost completed its production layout in the countries along the Belt and Road in Southeast Asia.

Simon believed that traditional business model had a role to play and must be preserved. However, the trend for the entire industry was shifting. Seeing the limitations of original equipment manufacturer (OEM) development at the beginning of his start-up and foreseeing that the cost advantage of the Pearl River Delta Area would eventually disappear, Simon planned ahead.

"It takes time to go from establishing an office to setting up a factory. The first step is to settle down. Once you get familiar with the local culture and build a connection with the local population, you can prepare for the next step: in-depth penetration."

Ronny Ho commented, "You can see that Simon has a whole picture. On one side, he expanded business through overseas acquisitions to drive growth with technology. On the other hand, enhance the production capacity in Southeast Asia.

The two tactics reinforced each other." Having worked with Simon over years, Ronny well understood his mentor's "expansion as well as defending" strategy. Once the direction had been set, the specific implementation was left to the team. Simon was not directly involved. This stance was in keeping with his "done by doing nothing" philosophy.

Simon was of the view that a boss should leave enough space and freedom for the team to play a full role. The most important matter is to give clear directions in order to achieve the expected results. He didn't visit most production sites in Southeast Asia until 2017.

In September 2017 during his first visit to the factory in Dhaka, the capital of Bangladesh, Simon was greeted like a star. Dressed in traditional costumes, staff representatives carried flowers and greeted him with a grand ceremony. When he left the factory, his audience of well-wishers refused to leave his side. "I've been working here for 10 years, but today is the first time saw my boss, very excited and did want to take a picture with him", said a veteran employee who joined the factory when it was first built. He waited in line patiently. When it was time to say goodbye, the employees spontaneously shouted in Bengali "fighting" and watched Simon's team leave.

Followed were his visits to factories in Jakarta, New Delhi and Colombo, which greatly boosted the morale at local factories.

Simon once had made a study of Chinese medicine, and he described the purpose of such visits as "taking pulse" and "boosting spirit" of the employees. After walking around the factories, he knew clearly what was done well and what needed improvement. Then, he would sit down with the local management, listening to their opinions and suggestions and then share his own views.

"Having a free hand in factory management yet caring for them and listening to them. It is like educating children. Teaching and guiding are important. Too much criticizing will lead to the opposite effect. The company is a big family, and every region is an important part." Simon Commented.

He requested the senior management at headquarters to make frequent visits to the sites, to understand the local operation. He demonstrated a caring and supportive approach and was, in particular, concemed about the challenges at the newly set-up factories. During his first visit to Cambodia, Simon heard that there was an imbalance in the ratio of men to women and hence, it was particularly difficult for men to start a family. After pointing to the areas in need of improvement at the

❖ *Wearing traditional costumes, staff in Sri Lanka welcomed Simon with the grandest ceremony.*

❖ *Staff in Bangladesh took photos with Simon.*

factory, Simon remarked in a half-joking yet serious manner: "Everyone here should work hard and make achievements so that he can at least afford to get married."

The managers, mostly unmarried men approaching their 30s, laughed at the pointed accuracy of Simon's joke.

Simon is indeed an amiable and a practical boss.

Starting in Hong Kong, the branches have expanded to cover over 30 countries and regions. Simon admitted that he had not been to every single one. "I visit Europe and America more often because the markets are there." After meeting with colleagues at the company, Simon would go shopping, which he believed the best way to understand the market. As a fashion accessory, labels are closely related to the retail market. "Go to a few big international brand stores. Taking a look at the flow of consumers and the goods quality, can effectively estimate the performance of the brand." From the observation of small details, Simon develops his grand plan.

From Simon's office, it was possible to see the runway of the old Kai Tak Airport, which first led to the Hong Kong' economy to take off and to visit, as well, the looming Victoria Peak unfolded in the distance. In daytime, scattered buildings on the opposite bank of river reflect the orderliness and efficiency of this cosmopolitan city, as a poetic scene of "clear mountains and flowing water". When night comes, sunset casts a rosy glow over the sky, and the shining lights and floating clouds deepen the picturesque aspect of the energetic scene.

In 2013, Simon moved the headquarters to Kwun Tong, which has come to be regarded in the history of Hong Kong as a place with "success genes". In the heyday of manufacturing industry, Kwun Tong accounted for one-fifth of Hong Kong's total economic output. The dense industrial buildings have become a home to many encouraging stories which best portray the "can do" spirit. In 2012, the government launched the "Energizing East Kowloon" project, aiming to transform this area into a new core business district and a hub for cultural heritage. By locating the new headquarters, which features the "Lion Rock Spirit", Simon expressed his commitment to "continuing the legend in the legendary place".

In the same year, Simon received the Ernst & Young Entrepreneur Award which was also known as "the Oscars of Business".

Since its inception in 1986, the Ernst & Young Entrepreneur of the Year Award, founded by Ernst & Young Accounting Firm, has honoured more than a thousand most successful and innovative entrepreneurs, including Michael Dell of Dell Computers, Howard Schultz of Starbucks Coffee, Jerry Yang of Yahoo!

and Jeff Bezos, the founder of Amazon. It is considered the greatest recognition for entrepreneurs who have built and run successful businesses with wisdom and perseverance. For this reason, the recipients of each year receive a great deal of media attention.

On the day of the award ceremony, Simon's schedule was packed: in the morning, he was interviewed by three media respectively from the Mainland, Hong Kong and overseas. At noon, he attended the press conference held by the organizer. In the afternoon, he joined in the group interview session. When appearing in the award ceremony later in the afternoon, Simon was still with great vigour. His eyes looked clear and determined. The huge posters of the winners were lined up in a circular pattern along the wall of the front hall. Walking down the red carpet, Simon stopped in front of his poster. Staring at the impression, he felt like "after all the sails, I still remain true to my aspiration".

Jiang Changjian, a professor at Fudan University, hosted the ceremony. He had also been the "Best Debater" of the first International Varsity Debate in 1993. After years of tempering, the "Best Debater" had become more sophisticated and mature in his speaking style. The impromptu question-and-answer session seemed to be mild, nevertheless, testing the responsiveness of the award winners. The interaction with the first few respondents did not go exactly smoothly. There was slight embarrassment in the grand atmosphere.

Simon's assistant was very nervous. He prepared a number of possible questions and would have preferred to rehearse, but Simon said no need. The grander the scene is, the more collected he becomes.

"Mr. Suen looks like a Confucian businessman," Professor Jiang phrased very politely in his opening.

"Not really, but I'm a little interested in traditional culture." Simon was quick to this short question. "What does traditional culture tell you about doing business?" The host immediately seized on the phrase, traditional culture. "'Sitting by a raging river, my heart is at peace, as the cloud above my head, at ease.' The two lines are my motto for doing business." Simon responded, speaking slowly each word in Mandarin. The host understood, recognizing the quote to be from Du Fu's poem *Jiang Ting (Pavilion besides the River)*. "Why?" the host went on to ask. "Emphasize the reality and recognize the changes." Simon answered. There was applause from the audience. The host looked at him approvingly and asked no further questions.

One minute on the stage, ten years of practice off the stage. For Simon, he never

❖ *In 2013, Simon received The Ernst & Young Entrepreneur of the Year Award, which is known as the "the Oscars of Business".*

❖ *Simon answered impromptu questions raised by the host in the awarding ceremony.*

stopped practicing. Over 30 years ago, a young man in a suit, who carried a shabby briefcase while braving the scorching heat and walking the streets to knock the doors for business, had never considered that one day he would stand on this stage, holding this trophy. The road had started from practical yet simple principle: do a good job. Simon had not failed to live up to his original aspiration.

"The label is my lifework and is part of my life. I've spent most of my time thinking about how to grow this company. That focus has never changed." When Simon is not smiling, he looks serious and formidable. The brave and ruthless boy living inside becomes faintly visible.

Sometimes, he shows an emotional side.

"I like to innovate and dream; it's the joy of life to see dreams come true," he raised his voice ever slightly, "but it's never easy to transform and move forward."

Simon has indeed moved forward from a business model focused on labels and tags before arriving at the RFID one-stop solution. He has also seen his business undergo significant changes until its ultimate transformation into SML. Simon led the expansion of SML from Hong Kong to Dongguan, Mainland China, then in Europe, America until the company achieved a global presence. This country boy had been through a great deal, even coming face to face with death. He came to Hong Kong with bare hands and attained prominence in the business world, creating a technology legend of a small label in the big world. The struggles, pains and hardship can be summarized with one sentence by Simon himself, "Being an entrepreneur, is not easy."

He continued, "But I've never given up. Senseless Simon Suen is such a senseless guy."

Simon didn't mind that his peer addressed him as "Senseless Simon Suen" when witnessing his persistence.

After receiving the award, interviewers would often ask about his understanding of entrepreneurship. In response to this frequent question, Simon wrote once late at night, "I often go down the mountain in the dark night. The mountains are as vast as the sea. Surrounded by shadows, I can barely see the path under my feet. Many times, in my life, I moved forward by feeling my way. However, no matter how difficult it might seem, by looking back and moving forward, we will be more empowered to create the future. Of course, we cannot become immersed in the past, instead, reflect on the past and turn the wisdom we have gained into a driving force to face the future challenges. This guiding principle forms the core

❖ *Simon attended the lunch meeting of the Ernst & Young Entrepreneurs.*

of entrepreneurship." Looking at the trophy in the corner of the conference room, Simon said: "I'm very happy to win the award, and more than that, I feel very comforted."

In 2007, while Simon was devoted to completing his business transformation from woven to print labels, American sculptor Richard McDonald launched his "Faith" series of works. Six years later, Simon saw the series in the sculptor's showroom in the States and immediately bought the No. 8 piece, titled "Blind Faith". What he liked most was the symbolic meaning of the sculpture: behind the blind faith lay persistence and perseverance.

The day after the award ceremony, the winners' news took up much space in major media outlets. Simon's statement that "The one who leads technology will dominate the market." was published and given prominence.

"Technology is not just a slogan, or only known by a few members of senior management. To the contrary, the entire team should clearly know, even the cleaning lady." Simon understood that knowledge and mindset of the team were keys to the smooth transformation. "The RFID business could not be run the same way as the traditional business."

Following the opening of the innovation centre in the UK, the company has set up technology hubs around the world to showcase its business capabilities to customers and to play the role of internal training centres. "We want everyone in the company to be technologically savvy and to know our products.

In the following years, innovation centres were established in the US, Spain, Germany, and Shanghai. At the same time, infrastructure projects to enhance production capacity in Europe and Asia were completed. The new plants in Turkey, Portugal and Myanmar started operations. By 2018, SML had built the world's largest RFID service bureau network, providing RFID technology service to the world's top fast-fashion brands. SML had managed to catch up with the competitor, which had a decade ahead start in RFID.

With the company growing bigger, Simon assigned himself a more singular task: just taking care of company direction. He left the team to be involved in the daily operations. But sometimes, on a spur, he would go to the factory to have a look: "Probably I am addicted to factories. It's fun for me." he laughed.

His friends and colleagues are all amazed at his energy. After attending an event in Beijing, he would take the high-speed train, visiting the factory in Qingdao. Afterwards, he went to look at his factory in Shanghai, then to Changzhou where he

❖ *Simon (middle) and the overseas team were in the first innovation centre in UK.*

made a stop at the joint venture factory before returning to Hong Kong.

It was Simon's first experience of taking a high-speed rail in Mainland China.

Back to 1991 when he went to Hanover, Germany, to participate in an industry exhibition, as the local hotels had been fully booked, he had to book a hotel in the nearby city Hamburg and took for the first time an express train, arriving each day by the ICE intercity to the exhibition. Jason Mak, who accompanied him, recalled Simon wondering when China would have such a high-speed railroad.

Simon bought a first-class ticket and drank his first-ever cup of coffee on a train speeding at nearly 300 kilometres per hour. The coffee cup by the window did not move. No drop fell out. The high-speed rail was fast and stable, and he expected the same for his RFID business. When SML was moving at the pace of a "high-speed rail", the company inevitably drew the attention of its competitors.

Simon recalled, "Our competitor was very nervous. In 2015, they formulated a strategy specifically addressed to SML, cutting prices of traditional products to edge into our market share and profits. Technology requires a heavy investment, and we are not a listed company, investing one dollar and raising the next nine dollars from the stock market. We have to pay all the ten dollars by ourselves".

"The strategy adopted by the competitor led in the short term to an adverse impact on the bottom line. But the prospect within the industry matters most." Simon did not mention the pressure that he was undergoing but instead gave an example. "Say, Amazon, how many people were optimistic about it? As an entrepreneur, you cannot always expect to be understood, which is actually hard to understand."

When the 2016 financial report was released, the net profit dropped to a low point. Simon spent a whole week meeting with senior and middle management where he disclosed the figures.

"We must tell our colleagues what is happening, what difficulties we are facing, and we need the support of the whole team to ride out the storm together."

Some management expressed reservations about this approach, thinking that financial data was sensitive, and the release might be risky. But Simon didn't agree, "You have to share with colleagues so that they will share your worries."

Simon has always been an advocate of transparency in the company's culture. He has a vivid metaphor, "The worst is to set up walls, which block all the possible exits and cause a traffic jam." In traditional Chinese medicine, it is believed that the cause of aches to the body can be traced to stagnation. A similar diagnosis can be applied to business management. It's better to be open.

Simon is empathetic, being able to view matters from the perspectives of his employees: "As a boss, you have to be candid. My colleagues could see that the company was not making much money, but the boss still wanted to keep the company going. It's a bond of affection for the organization as well as his sense of responsibility to the several thousands of employees and their families."

Simon said emotionally, tapping his fingers against a table. "The concerns of the management could be understood. I'm sure that some staff felt that it's none of their business, but most of them would understand our difficulties and have much to contribute in finding solutions." For several years, Simon just gave advice on the general direction and did not engage in daily operations, but this time, he said, "I must take the lead." Cherry Cheung, one of the senior management team, has witnessed his charisma during her many years working with Simon. "The colleagues trust him. He is a spiritual leader. When he talks, the colleagues feel secure. At that time, Dong Hing's earnings were very good, except for the year of 2008 when there was a financial crisis. During that crisis, Simon practiced a similar form of candor. Knowing the difficulties, his employees were willing to take the cut in pay to help the company through the challenging period."

Cherry added: "He never tries to fool his employees. In the following year, when the situation improved, he gave his employees a pay raise. Compared with foreign and other local Hong Kong companies, our company exerts humane touch. The boss shares happiness as well as the difficulties with the employees." Money can never gain the employees' commitment. One can only achieve a long-lasting bond through the heart.

Despite the unsatisfactory profitability, Simon decided to allocate a sum of money to renovate office buildings at Dong Hing. When the proposal was brought for the first time to the Board of Directors for discussion, management was, by and large, unsupportive. Labour costs had risen in the Pearl River Delta area, and orders had begun to shift to Southeast Asia. These events, coupled with an uncertain global economy, had led understandably to management, in general, to be conservative. The prevailing attitude was better to keep everything unchanged.

But Simon didn't agree.

A series of policies supporting the transformation and upgrading of traditional manufacturing industries and encouraging enterprises to innovate independently have been introduced by the government. The plan of the Greater Bay Area, including Guangdong, Hong Kong and Macau, was to build a world-class advanced manufacturing cluster. Shenzhen and Dongguan would serve as a hub. He recognized immediately that to thrive in such an environment, Dong Hing would have to modernize its infrastructure. The first step was to renovate the offices: necessities if the company was to recruit a talented force.

"It's not that I hadn't struggled." Simon didn't try to conceal his hesitation. He said, "Large sums of money are required. I know better than others that cash is king. However, if we don't change in response to changing conditions, we won't have room for further development and might even struggle to survive. Time-related costs also need to be factored in. We'd still have to face the choice after one or two years. The situation might become clearer or possibly more uncertain in the future. Investing always involves risk. What we need is courage and prudence."

In the face of the opponent' advance and under the tremendous pressure, Simon calmly tackled the battle.

It's not easy to be a pioneer. For a while, whenever pondering the changes that must occur, he couldn't fall asleep. Thoughts, dreams and blueprints came over him like ocean waves. When closing his eyes, Simon remained fully occupied with possible sources of anxiety and feeling of uncertainty. He could only sleep with the help of sleeping pills about three hours each day. He is so attached to the company, but innovation and transformation means risk. The struggle is inevitable.

If he had wished, Simon could have retired with a large sum of money. Simon had kept locked a letter of intent given by a party intending to acquire the company at an attractive price. It's not, then, as if he hadn't considered leaving the business. During the price war initiated by his rival and the initiative to develop the technology business, Simon became physically and mentally exhausted.

But he didn't take it. A thin and small RFID tag held in Simon's black leather-bound notepad has an answer. It's a tag which Simon had picked up during his invariable "shop tour" in the overseas business trip; a brand which is the company's first RFID customer.

As an impulsive teenager, when Simon planted rice seedlings and looked at the sky in the water, he determined that he "can't just be a dreamer". Today, as a global entrepreneur who believes in the technology innovation, Simon has realized the spirit behind the statement: "in doing business, you must have the fighting spirit". It is his challenge and his mission to run well SML. The concept of a big world in the small label demonstrates his large ambition, profound wisdom and broad vision.

The clouds dispersed, and Simon became determined.

In 2018, SML was awarded the DHL/SCMP Hong Kong Business Award in recognition of its outstanding achievements in overseas expansion as a Hong Kong company. The local Business Award ceremony was first held in 1990 to recognize outstanding local entrepreneurs and enterprises and to promote the entrepreneurial spirit of local. Famous businessmen, such as Li Ka Shing and Dr. Sir Gordon Wu Ying Sheung, have been honoured with this award. Simon's acceptance speech was simple. He gave three thanks to firstly, the organizer; secondly, his entire SML team; thirdly, his family. But Simon's achievements are most of all a product of his own perserverance and persistence without which the success of SML would not be possible.

Before the Chinese New Year in 2019, Simon flew to Osaka, Japan. The latitude in Osaka was high. The winter was cold, but there was plenty of sunshine. Wearing a blue-checked beret and a black safari-style cotton jacket, Simon looked fashionable. Standing on the street, he took a photo. Simon was squinting and smiling. Seeing the photo, his friends thought Simon looked like a film director. The meeting with one of Japan's top five trading companies was scheduled for the afternoon. In the morning, he was free to wander the busiest street in the city where he bought a red trench coat

❖ *In 2018, SML was awarded DHL/SCMP Hong Kong Business Award.*

for his daughter. She had joined the company the previous year. He said, "Red goes well with her."

From his many battles with competitors, Simon arrived at a solution. The transformation of the company required new thought and guidance. It was high time to introduce strategic partners who might give impetus to the company to realize the concept of "consumer technology".

The company had found for its first partner, a Mainland technology developer with Inlay R&D capabilities. For the second partner, Simon targeted both investment funds as well as prominent companies that could spark the development of SML. Before flying to Japan, Simon met with his team to understand the progress of the project: "The partnership is not just about raising capital. The market effect brought by the partnership itself is more significant." A photograph is set on a low built cabinet behind Simon. It shows Simon at an event. The photographer captured his look just right. Simon strode into the venue with the vigour and poise of a man who "walks leisurely regardless of the storms and waves".

After his morning shopping, Simon changed into a beige gray plaid suit with a red and blue striped tie, to visit the company's headquarters. He looked spirited, ready for a final communication on key issues between the decision makers from both sides. Simon had not participated in the first few rounds, leaving the team in charge and only joined for the final decision.

The relatively straightforward part of the negotiations had occurred before the meeting. The bottom-line issues remained, and each side became very adamant. The negotiations from this point forward were bound to be difficult. Simon was skilled in digesting information and not in a hurry to state his stance. He reminded his team members to be observant and prudent.

"The partnership must be seen as a win-win by both sides. Each party must benefit from cooperation. Only by doing so, will the partnership be long-lasting and solid." Simon took a longer view. "Loss does not necessarily mean loss. Future synergies are what will lead to long-term growth."

Simon envisioned that the company would become listed as a consumer technology company in three years. Both parties would stand to benefit from this move. The representative of the Corporation immediately understood the need to work together in harmony to achieve a win-win situation. He came over and shook Simon's hand.

That evening, a sumptuous Japanese dinner was arranged. Simon made an

❖ *Simon and the Executive Director (left)*

exception and drank a little to celebrate a collabouration that had grown out of his wisdom and resolution.

On a mildly chilly morning in October, Simon stood in front of the picture window, extending from the floor to the ceiling of his office, gazing on the haze hovering over the sea. Two of the three major international rating agencies had downgraded their outlook on Hong Kong from "stable" to "negative", pointing to the increased uncertainty in the environment. Simon had the vague feeling about the project with Abu Dhabi Investment Authority (ADIA), the world's fifth largest sovereign wealth fund, which underway for the past several months, would be concluded on the day.

He held a short meeting with the core members of the project team telling them to be prepared for failure, but never lose heart. "It's like in horse racing. Though our company has great potential to win, the risk of losing might be also high. During stressful times, Simon had been able thoughtfully to reduce the pressure on the team.

It was 1 PM Hong Kong time and 7 AM in Abu Dhabi. Simon's cell phone vibrated, and the screen showed that a new message had been received. After reading the note, he forwarded the content to the project team. Following six months of negotiations, the project authorized earlier that September by the Fund Investment Committee failed to gain the final approval of the Investment Decision Making Committee of ADIA.

Immediately afterwards, he convened a special meeting to brief key management on the unfavourable result. As usual, Simon practiced the transparent management style. The second item on the agenda was to plan for next year and continue with the original plan to achieve IPO in the period from 2022 to 2023. Simon understood that the team might have been a bit down. The project had appeared to be on the verge of success. "Of course, we are disappointed, but it's important not to look at the matter in such a grim way. After summarizing lessons and experience, we're ready to set off for a new journey." Simon commented slowly and resolutely, his eyes sparkling.

However, the road was more bumpy, rugged and difficult than expected.

Though the ADIA project was unsuccessful, Simon did not give up on the idea of bringing in a fund. He accepted proposal of Citi Bank to select a fund partner through bidding. Surprisingly, the market showed strong interest in the bid, with more than 40 funds from around the world bidding for the project. For a week, the project team met with each fund, and after the first and second round of selection, four funds were selected as finalists.

In January 2020, Bloomberg was the first to report that leading global funds were vying for a stake in SML. The news was immediately reprinted by many financial media outlets. When the economy was weak, the news of a label company being favoured by funds was no doubt eye-catching. Meanwhile, the news on the outbreak of Covid-19 in the Mainland had not yet gained much attention. But soon after, the pandemic garnered the headline, and indeed, its impact on the business climate was immediate. Global stock markets plunged. In the US, there were four unprecedented meltdowns within 10 days. The turmoil in the financial market cast a shadow over the global economy. The four candidate funds, without exception, requested to postpone the project, which had been progressing smoothly, once again took a 180-degree turn for the worse.

The impact of the pandemic has made layoffs and cost-cutting a top priority for many industries and companies. As a leader with a gift for facing a crisis, Simon saw earlier than others that some companies may make profit while others may lose more in this pandemic. As such, resources need to be kept for future development.

He attended the management meeting, which he seldom did before and said, "Timing is not on our side, need to get ready to strive for the next round", encouraging the management to stay calm and quickly implement countermeasures to minimize the impact of the pandemic on the business.

"One's personality can be best revealed in his responses to change. To seize the opportunity in uncertain period where people must fight against apparent fate and desperate rivals, we have to bet on our vision, rely on our courage and live by our wisdom. The one who can meet that challenge is the hero!" Simon explained. In Simon's view, surviving the turbulence and withstanding the test of time are solid proof of the company's value. Despite the sudden impact, he was confident with his principle: "Emphasize the reality and recognize the changes".

Simon's confidence is not unreasonable. The situation began to ease as the pandemic gradually came under control. An investment team jointly made up of two overseas funds extended an olive branch and decided to take a stake in SML: no doubt exciting news amid a market downturn. The project soon entered the practical phase. Besides, SML's own performance has also exceeded expectations. The overall performance in June was even better than it had been in 2019. Everything seemed to be on the upside.

However, no one expected that as the second and even third waves of the pandemic began to hit the world, countries once again locked down in order to combat the outbreak. The impact of the economic shutdown on the retail sectors and other consumer goods industries has been profound, and the effects of the weakened economy are still surfacing. In the last conference call between the two parties, one of the funds acknowledged that it was difficult to raise money, which became the straw that finally broke the camel's back. The introduction of the funds project once again was set back.

"It's said that three twists and turns will lead to a satisfactory result, but for us, probably four or five twists and turns won't work." Despite the successive changes, Simon could still joke about their situation, never looking frustrated. "Each time it is a step forward, but in an S-shape. Although there's a detour, we are still approaching our goal. This has never changed."

After several days of hot weather warning, strong winds blew into Hong Kong. The Observatory signaled the prospect of a No. 3 typhoon without signaling No. 1, which was rather uncommon. On the early morning of Saturday, the wind became stronger, the trees swayed, and the rain began to fall. As usual, Simon started his morning routine: swimming before practicing Ba-gua walking. Then he had breakfast. Reading newspaper was also an important morning exercise for Simon. Afterwards, he went to the office, browsing more information. The whole set of routine had been kept for many years. For Simon, the best way to navigate a crisis is to seize the certainty in an uncertain world. In the context of the once-in-a-century pandemic and the changing world, Simon's calmness in the face of adversity is extremely valuable.

In the afternoon, after once again practicing Ba-gua Zhang, his martial arts, Simon recalled the challenges in becoming an accomplished athlete: "I was not tall, didn't really have an advantage in athletics. Normally, this would not be the pursuit. But I didn't have a choice. It was my only chance." He laughed as he made that remark: "Instead of complaining about the cards in hand, we should cherish the hand we are given. The greatest ability is to look at each game calmly and play each card well. Respond decisively to crises and get out of the situation quickly. Even in the most difficult times, there are still some enterprises gritting their teeth and laying the groundwork for the future, even growing rapidly despite the unfavourable market. No matter how hard and difficult it will be, we must go forward courageously. It's well-known that the dangerous peak offers the most beautiful view!"

There are many outstanding entrepreneurs in the business world, but rare real legends. Speaking of the big events of the times, Simon has experienced a great many from labour reform, the opening-up to the return of Hong Kong and the financial crisis. Perhaps it is his rich experience and self-discipline that have enabled him to stand proudly at the head of the tide, to move steadily forward, and to become better amid the storm. In the mighty torrent of history, individuals are as trivial as dust. Only those unshakeable beliefs and practices come to be like the light in the cracks of time, bringing inner certainty and vitality as well as abundance and hope.

Simon continued, "Technology leads the future. Only by walking side by side with change can enterprises sustain growth. SML will be a listed company that provides consumer technology solutions and has equity participation. As its founder, I have taken the company on this new voyage, and I believe that the future management team will do what it takes to carry on this philosophy and mission. Remembering the reasons why we started this journey will enable our accomplishing the vision. In a few years, we will look back on 2020 as the most extraordinary and memorable year in our lives."

The years are fleeting, and the world is changeable. From a regional label manufacturer to a consumer technology solution provider, then towards becoming in the near future a listed company, no one knows better than Simon the character and fate of his company. Starting an enterprise can be difficult undertaking. Its transformation into a global leader is yet more challenging. But turbulence and uncertainty are a part of any journey worth making: what is the most important for the company is to stay firm in the storm. With perseverance and readiness, SML is destined to write a new legend.

In 1980, the founder of SML came by water, riding over the waves and tides. Over the past forty years, Simon has witnessed how the struggles against the hardship have become inscribed in a business legend.

Coming by water, riding over the tides of 1980;

Legend of label, forged by 40 years of hardship.

# LEGEND

/

## A UNIQUE AND SINGULAR BIOGRAPHY

A legend is about an individual standing apart from the crowd. A singular individual possesses superior ambition, courage, thought, talent, and wisdom; the singular individual holds dear honesty, morality, loyalty, ethics, and faith. That's the Legend of Simon Suen.

CHAPTER

# VI

# INTEGRITY

/

## CULTIVATING ONESELF

# Sincerity: Carried by Action, Realized by Persistence

It was six o'clock in the morning. The sky was not fully bright. At London Heathrow's Terminal 3, a CX255 flight from Hong Kong had just landed. After 13 hours on a long-distance flight, all the passengers stepping out of the cabin appeared somewhat weary — except for one. Wearing gray sweatpants and black sneakers, Simon pushed a small suitcase with a black briefcase slung on the trolley, flying down the passageway in a manner that was particularly eye-catching in a crowd of somnolent travelers. Simon walked lightly and fast, landing on his toes, like the Qing Kung master of Louis Cha's writing. His companions tried to keep up with his pace. However, even though they were carrying only small bags, they could not catch up. When his entourage arrived at immigration, they were panting; their faces flushed. Simon, however, looked rather relaxed, saying that the stroll had been a shorter distance than his usual morning walk. He had to exert a faster pace in order to achieve the same intensity typical of his morning workout.

The Confucian way to cultivate the self is composed of determination, studiousness, and introspection, as well as reforming one's negative habits, and practicing what one preaches. In Simon's view, the five aspects are integrated, and putting into practice what one preaches remains his core principle. Simon explained, "You see, determination, studiousness, introspection and reforming are all related to practice. In *The Analects of Confucius*, it's written: 'A gentleman takes it as a disgrace when his words outstrip his deeds'. In other words, don't be a bragger. Do it." Simon's practicing Ba-gua Zhang, came to exemplify his beliefs. In his sixties, Simon was still able to practice the Ba-gua Zhang whenever he wanted. His waist was flexible; the movements of his palms exuded a smoothness and power while his body remained firmly unmoved. The imposing manner of his Ba-Gua Zhang was obviously the result of years of persistent practice.

In 1983, the White Swan Hotel, located at the southern end of Shamian Island in the city of Guangzhou, facing the White Swan Pond, was officially open to the public. As the first large-scale modern hotel designed, constructed, and managed by Chinese, the White Swan Hotel was a successful seed sown by the country during the period of reform and opening up. In November 1984, the signing ceremony of the joint venture contract between Nanwei Electronics Industry Company and Nanwei Electronics Investment Co., Ltd. was held at the White Swan hotel that was universally acknowledged to be a window of the times. Liang Lingguang, the then Governor of Guangdong Province, and Pan Jiefu, the First Secretary of the British Embassy to China, attended the signing ceremony.

The ceremony was held in the banquet hall, where a long table covered with golden tablecloths had been laid out complete with the documents awaiting signature. The representatives of both sides were seated in the front row. The guests in the back row lined a semicircle. Most wore western suits, a few in Mao suits. There was a painting of rolling hills setting on one wall alongside a red banner with the words "Contract Signing Ceremony". A huge round table sufficient to seat 20 guests had been situated in front of the signing table, indicating a grand banquet would be held afterwards. On the table was an exquisite flower arrangement whose inner ring contained a two-tiered tower-style bouquet, the periphery decorated with garlands; the vermilion Chinese highchairs being placed next to the table. Each chair had a hollowed-out carving etched into its back. In the distance, the cameramen recorded the ornate proceeding. The scene was grand and magnificent.

In Simon's house in Shangsha village, there is a photo of that signing ceremony. "This is my father", Simon pointed to the man on the far-left side of the long table, who was taking a pen from his inside suit pocket to sign the document.

"My father was a trader. During the reform and opening-up period, he facilitated a number of foreign-invested enterprises to cooperate with companies from Mainland China." When Simon talked about his father's past successes, his voice grew filled with admiration and pride. "He has a knack for dealing with people and I may've inherited it."

Simon, who started out as a broker, does not deny his genes for business that must have come from his father.

In Simon's eyes, his father had a free and easy spirit. He was generous and willing to spend money to help others. By doing so, he made many friends. This kind of personality was obviously advantageous in the trade business. But when he first saw his father "generously" tip a taxi driver 100 HKD, Simon couldn't understand the practice and even became a little annoyed. At the time, Simon had only a monthly salary of 600 HKD. One hundred HKD was undoubtedly a huge sum. His father's company eventually became shuttered. Having observed the difficult event, Simon frequently reflected on the lesson that could be gained from his father's

❖ *In 1984, Suen Shing attended the signing ceremony of the joint venture contract between Nanwei Electronics Industry Company and Nanwei Electronics Investment Co., Ltd.*

ordeal.

"In fact, he was very forward-looking, but a person's success has a lot to do with opportunities and financial resources. He played the role of a trade consultant and had to balance the interests of both parties. Sometimes it was just a thankless task. His personality has naturally influenced my own development, and it's fine to adopt the strengths of his character, but more important to see the weaknesses and find the right way to overcome these deficiencies. The best practice is to face up to one's own shortcomings and correct them; these twin steps are important parts of self-reflection and cultivation." Simon jokingly said. "My challenge is to avoid the flashy manner in my gene. To do business in industry, I must stand on solid ground."

As an industrialist, Simon sometimes also "spent money like water". However, the purpose was to buy machinery and equipment as well as improve the living environment of employees. In the 1980s, he never hesitated to buy equipment sometimes costing more than one million RMB. In addition, the staff dormitories were equipped with air conditioners. In those days, having an electric fan at home meant a lot in the countryside. The installation of air conditioners in factory dormitories naturally attracted a lot of attention. Simon remarked and laughed, "At the very least, one air-conditioner motor was stolen each day and sold for parts."

Simon had seen firsthand the effects when he was opening factories in the countryside. During this period, Chen Manlin, his colleague from Chang'an Middle School, joined Dong Hing. Having just returned to Shangsha from Xinjiang, Chen had nothing to do. Fearing that he would go astray, Simon gave him a position at Dong Hing, where he was in charge of security. As a graduate of a physical education school, Chen was tall and sturdy. When standing in front of the gate of the factory, he appeared a bit of an imposing figure. After he took over that post, not surprisingly, the motor thefts stopped.

In stark contrast to his generous investments in the factory, Simon remained stingy in rewarding himself.

Chen Manlin recalled, "At that time, when he hurried back after a few hours' bumpy ride to Dong Hing from Hong Kong, it was usually between 8 and 9 PM. The canteen had been closed. Simon, though, had not had dinner. He ordered three dishes at a small restaurant next to the factory: a crucian shorter than a piece of chopstick as well as a bowl of steamed eggs and a plate of vegetables. When a crucian was not available, he ordered a grass carp. The dinner for all four of us — Simon, Uncle He, Uncle Jing and me — cost 70 RMB." Chen observed for years after, Simon dined on similar fare, each person's serving less than 20 RMB. The habit did not vary even as Dong Hing grew to possess a workforce of 700 and though the company was making a great deal. Its annual sales exceeded 100 million RMB, with a double-digit net profit margin.

"I think it's great. The meals provided with all the essentials: fish, eggs, and vegetables," Simon remarked. Having suffered a lot since his childhood, he had grown used to living a simple life.

When it comes to eating, Simon has several characteristics.

First, he eats fast. The person next to him may be just halfway of the meal, but Simon has already finished. This may have something to do with his character. He is quick-tempered, wanting to finish every task as fast as possible. Eating is no exception.

Secondly, he doesn't care about the content of his meals, as long as he isn't starving. That's why he never thought about changing the menu when same dishes had been served for many years.

Third, he is not picky about the environment where he dines. Usually, he ate in the same small restaurant next to the factory. The restaurant owner knew Simon so well that the owner didn't need even trouble taking the order from his longtime regular. When talking over meal with his colleagues, Simon would slap the table whenever he became excited. Those who ate regularly with him had grown accustomed to the habit. If others were in his company for the first time, he would explain apologetically that no one taught him table manners when he was a child. Now it had become a habit that could hardly be changed.

Simon is not picky about clothes either. When forming a close relationship with Young Law, the boss of Laws Textile Industrial Ltd., he often wore Bossini, a popular brand of Laws. Simon said, "100 HKD is enough to dress me from head to toe. It's not something intentionally, just feel right to do so, being frugal."

A colleague from the administration department collected old photos for a special issue celebrating the company's 30th anniversary and found one that Simon wearing a white Bossini T-shirt and beige cloth pants, received a Swiss loom supplier.

Simon recalled that when registered for his marriage certificate, he even didn't have any decent clothes, having to borrow suits, ties, and even shoes from his father. "I had to change my clothes right after signing the certificate, to deliver the labels, and the formal clothes were obviously unsuitable. That evening after the busy day at work, I invited a few fellows also in Hong Kong out for a meal. It was like a wedding feast." With hard work and frugality, soon after he started his business, Simon bought a 500-square-foot condo in City One, Shatin. Although I was able to buy a condo, I couldn't afford to decorate it. The floor was uneven, and I had to use a brick to even out the slope."

Once he was recognized in the subway by a former student who he used to teach in his hometown. Simon invariably took public transport.

Only in the last few years has he started to take a business class. His children have grown up and are concerned about him, and they don't want him to save money in this way. Simon said, "I used to take the 'toilet seat'. As soon as I got on the plane, I would automatically turn right and enter the economy class, searching for my seat, usually the one nearest the toilet." He jokingly called it "toilet seat". Being quite action motivated, Simon did what he wanted and would take many spontaneous joumeys. He booked the cheapest tickets on the fly, so the location of his seat was usually awful.

Simon said, "The company has a travel policy, allowing employees of certain level as well having a specified flight time to fly business class. I didn't follow this rule although I met the criteria." Talking about his longstanding "violation" of policy, Simon sounded full of pride.

Once he was able to delegate responsibility, Simon began to take some time every year to travel with his friends for relaxation. In 2018, he went to Tokyo. When doing business, he is a man of action, able to visit four cities within merely two days. However, he did manage to adjust to the different lifestyle, slowing down enough to enjoy life. He wandered around the Aoyama Street of Roppongi for a few hours, browsing antique shops and art galleries. When feeling hungry, he found a Japanese ramen restaurant and bought a bowl of ramen for 750 yen from the automatic ordering machine. Then he sat down at the bar opposite the open kitchen, watching with interest the chef making by hand the noodles before cooking each strand and serving them in a steaming hot soup. The quiet waiting process was enjoyable for him.

Simon also started to develop some theories about eating. When ordering dishes, Simon was willing to try those that he had never eaten before. It's a reflection of his boldness. He became fairly fond of a small restaurant that he had found accidentally in Tokyo, having dinners there two days in a row. The restaurant was easy to miss. It was located on the second floor of a building at a three-way intersection in Shinjuku. It had limited seating, just a round bar and four tables. After walking around all day, the moment his friends and he sat down, they started feeling hungry. All could not wait to eat the salted appetizers except Simon.

He waited patiently for the main course. The salt-grilled fish head was served very late. Its blackened surface did not, at first glance, appear to forecast a tasty meal. He took a look, "Not necessarily bad". As Simon had expected, the dish actually was very delicious, and everyone liked it. The next day, they went to the same restaurant again and ordered the same fish head, but the dish had already been sold out. His friends were a bit upset, so they chose another kind of fish head, requesting the dish be cooked the same way. After tasting it, he said: "This fish has its own particular flavor." His attitude towards food reflected Simon's own optimism as well as openness to possibilities. Simon's choice not to partake in the appetizer grew from his clear sightedness about priorities. In Simon's words, "The stomach has limited capacity. When it is stuffed with appetizers, there is no room to enjoy the main course." In small matters such as eating, he often has such epiphanies. It may appear accidental but actually is the result of years of self-cultivation.

Twenty years ago, this kind of "slow life" would have been unimaginable for Simon. He said, "I'm quick-tempered and impulsive."

In addition, he is a little bit generous and is always willing to help his friends. Such combination can lead to many incredible results. One incident from the 1990s shows his affection for and trust in his friends while also proving his impulsiveness and quick temper.

In the 1990s, cell phones were not popular amongst the general public. Primarily, businessmen used the device. Because the similarity of the proportions of a cell phone to those of a bottle, it was also jokingly called a "bottle". One day, when Simon was with Raymond Leung, suddenly the "bottle" hanging from Simon's waist rang. Leung had been his customer when Simon started up. Gradually they grew to be good friends and often travelled together.

Simon answered the phone. The voice on the phone was hardly audible. Faintly, Simon heard "Can you hear my voice?"

"Yes, yes. Is it A-Gao?" Simon thought the voice on the phone sounded like his old friend A-Gao.

"Yes, yes, it's A-Gao. I'm in trouble in Mainland China and need money urgently. Can you transfer some money now? It would really help."

Simon has a wide range of friends both inside and outside the industry. They also worked hard, and whenever one encountered a hardship, the others tried to help if they had the means. Simon said, "Okay, okay, give me an account number." After the call, he immediately called his old friend, David Lai, in Shenzhen to arrange money transfer. Raymond was suspicious and muttered, "Has he gone to the Mainland? Probably still in Hong Kong? I saw him in a restaurant a few days ago... Would you like to call him again and ask?"

Simon waved his hand and said, "Don't bother. I can recognize his voice."

There was a limit for a single remittance. David called Simon after the first remittance and asked whether he'd like to continue with the second. Simon opened his mouth and wanted to say yes, but Raymond suggested making a phone call first, asking if A-Gao had received the money. Having known Simon for many years, Leung understood his temperament. A mere "no" wouldn't stop Simon. He found another way to be heard.

This time Simon did take Raymond's advice. He made a phone call to A-Gao.

The phone was connected, and the first sentence Simon said was: "The money was just transferred to you. Have you received it yet?"

"What money?" A-Gao was confused.

"Aren't you in the Mainland?"

"No, I'm in Hong Kong."

Simon immediately realized that he had been deceived. Naturally, the transferred money could not be recovered. Touched by his loyal friend, A-Gao offered to pay Simon back. Simon, of course, refused, saying that he bought a lesson for his impulsiveness.

Simon said that he was a factory guy who was often impatient and quick to lose his temper. When he started to manage Dong Hing, he would grow angry when he found that some workshops couldn't follow "5S system", a standard process to sort, standardize and sustain production levels. He would often lose his temper if the quality was not up to standard or when the pattern made did not meet customer's specifications. In one case, the labels deviated from colour specifications caused by inexperienced workers. The customer had to be compensated with a relatively large amount for the error. The Dong Hing team held a meeting to review the incident.

Seeing that the department head still did not realize the seriousness of the issue, Simon became furious. The sound of his banging on the table reverberated beyond the walls of the meeting room. "Love well, whip well." Behind his stormy anger was his aspiration for perfection. At the end of the meeting, he talked to the department head alone. After having been severely criticized and was near the brink of tears, the head was frustrated when entering Simon's office, but upon leaving, became totally convinced as to the importance of correcting the error.

Sitting alone in his office, Simon saw the lychee trees that had been planted when the factory was built had already flourished. The green leaves were full of vibrancy. At that moment, he realized that he needed to learn to calm down.

"Flying into a temper might work when I managed a factory, but it won't when the business became bigger. Losing one's temper won't solve the problems," Simon commented.

Simon had already worked very hard when starting his business, and the resulting stress led to pain in his shoulders and neck. At that time, he was not yet 40 years old and had already seen many doctors, but the problem persisted. A friend suggested that he learn Ba-gua Zhang, which was a good way with multiple benefits.

"I always believe that health is one of the keys to maintaining and growing successful business. Practicing Ba-gua Zhang allow one to develop will power, determination and concentration. In addition, it can also help its practitioner to calm down." Simon said. Likewise, his performance of Ba-gua Zhang and Chinese Kung Fu coincided with his fighting spirit.

Having set upon a course, he began to search for a Kung-Fu master. Soon he was introduced to Li Dehua, a master of Ba-gua Zhang. Li wore shoulder-length hair with a wisp of white in the middle. He was sturdy, dressing often in a traditional Chinese-style overcoat with coiled buttons. All these attributes fit with the image of a martial artist. Soon after meeting Li, Simon decided to emulate the master, even following Li's style to let his hair grow a bit longer. As an expression of honour, Simon held a dinner for apprenticeship.

The dinner was held in the Victoria City Restaurant in the Sun Hung Kai Centre, which was completed in 1981 and is located at No. 30 Harbour Road, Wan Chai. The 53-floor building facing the Victoria Harbour was the first skyscraper in Hong Kong. In 1982, Chung Kam, known as the "Godfather of the Chef", rented a shop on the second floor of the Centre and opened Victoria City Restaurant specializing in upscale, new and authentic Cantonese cuisine. It quickly became the dining hall of the rich and the celebrities and has received many dignitaries, winning many international accolades. The French newspaper, *International Herald Tribune*, named it the fourth best restaurant in the world, and *The New York Times* called it "the most authentic Chinese restaurant". Simon is not extravagant, but in important matters, he values and emphasizes the sense of ritual. He booked a private room in Victoria City for the dinner. His father Suen Shing, his wife Mary as well as his good friend Wu Yan Wah, and Master Kwan Tak-hing and Kwan's son all attended the occasion.

Master Kwan Tak-hing was a popular Cantonese movie star in Hong Kong in the 1980s and 1990s. He starred as Wong Fei Hong, a martial arts master, in a series of films that deeply impressed their audiences. Master Kwan had been committed martial artist for many years. In his early years, he learned Hung Kuen, the great fist of Southern China, a discipline belonging to the White Crane School. Besides, he also originated the "Omni-Directional Gang Rou Fist" in 1970s. No matter in film or in his own practice, Master Kwan remains a true martial artist.

The private dining room in the Victoria City was very spacious, with a four-panel screen separating tea and dining. The chairs were gleamingly black, set with a long table inlaid with marbles. A golden wall clock was arranged on the wall; underneath the time piece, a watercolour depicted a moon beside the river. All these decorations presented the classic style and the elegance of an old-fashioned

Cantonese restaurant. Li Dehua performed Ba-gua Quan in front of the guests. Quan means fist, and Master Kwan nodded in recognition of the artist's skill, smiling while not saying much.

Halfway through the meal, Master Kwan said to Simon, who was sitting next to him, seemingly inadvertent but intentional, "The boxing is good, but his Chi is not as good as yours." Simon was surprised when hearing the comment. He was a beginner, but Li had been practicing martial arts for many years. It would only be reasonable that the apprentice was inferior to the master. How can it be the other way around? Later, he gradually understood that the Chi mentioned by Master Kwan referred to internal strength.

The cultivation of internal strength is related to personal virtue. To practice martial arts, one should have martial virtue. Only a righteous mind can lead to the successful practice of martial arts, so righteousness is the tenet that martial arts practitioners should uphold. Although Li was proficient in Ba-gua Quan, he liked alcohol very much, which affected his cultivation of internal strength. As an old school martial artist, Master Kwan had sharp eyes and recognized the strong internal strength of Simon. On that day, Master Kwan wrote calligraphy which said, in essence: "Sincerity can make a business. Filial piety means to listen to parents' advice. Think twice before leaping. Endurance leads to success". He emphasized to Simon that the cultivation of virtue should go ahead of martial arts practicing.

After the dinner, Simon took Li with him wherever he went and used every unoccupied minute to learn Ba-gua Quan. At that time, his company was beginning to go global and he often had to fly around the world. In order to seize the time to learn martial arts, he even took Li with him when he travelled overseas. According to Ronny Ho, often on business trips with him at that time, Simon and Li would practice Ba-gua Quan in almost any setting. On a whim, they would practice "snake-shaped hands" in the middle of eating, or practice eagle-claw fist on their way to some place. Simon made great progress. His entire body became much sturdier. Since then, he started to wear his hair long, emulating the style of a martial artist.

In 1998, Hong Kong Kung Fu superstar, Jackie Chan, made his debut in Hollywood with his movie *Rush Hour*. Foreigners started to become acquainted with movie stars from the region. Jackie Chan also had a slightly long hairstyle. Simon appeared, at first glance, to resemble the star. "Once we went to the United States. When queuing up at the airport to go through immigration procedures, the foreigners in the next line pointed at him and said 'Jackie Chan, Jackie Chan'. They looked

❖ *Master Kwan wrote calligraphy to Simon, which said, "Sincerity can make a business. Filial piety means to listen to parents' advice. Think twice before leaping. Endurance leads to success."*

❖ *Photo taken on the street of San Francisco. From left: Simon, Li Dehua and Ronny Ho*

very excited," Raymond Leung, who often travelled with him, recalled, and said that such things actually often happened.

Simon said, "In the first few years, like all beginners, I was always curious about my progress. What was my level? I wanted to find someone to compete with." He believed that if one wanted to compete, he/she must compete with the best. That opponent could drive the emerging martial artist to become successful. Li Dehua had a lot of friends in the martial arts circle. He introduced Simon to the national Drunken Fist and Sanda champions at that time so that Simon could learn from both about martial arts. Simon had an album, containing many photos taken with fellow martial artists. He pointed to one photo where the artist appeared to just a big boy, and said, "Don't underestimate him. He used to be the champion of Sanda and was very skilled in fisting and kicking." Simon sounded full of admiration.

"It's a crucial stage. If you don't compete with the best and don't get beaten, how can you know your own weakness? How can you reflect from the failure and improve yourself? If you just compete with an ordinary opponent, it's meaningless to win." Simon said. Martial arts practice is more than a physical exercise. It is a mental and spiritual discipline."

Matt, a Japanese, was in charge of UPDL, had been a partner with Simon's

❖ *Photo taken in Australia in 1999. At first sight, Simon indeed looks somewhat like Jackie Chan.*

❖ *Photo taken in Japan in 2018: the styles of the martial artist*

company. He had been practicing Kendo for many years. In addition to working with Simon in business, he was also proud of the role that martial arts played in their deepening friendship. There were a number of Kung Fu enthusiasts on Matt's team. In the 1990s, Simon took Li to visit the US headquarters of UPDL. When those blond colleagues heard that Li was a martial artist, they showed great interest and asked him to demonstrate on the spot. Standing next to Simon, Li took off his suit. At a lightning speed, he darted out, grabbing the right hand of one foreign colleague while the same time, feigning a punch with his left hand. Li's swift and agile moves impressed the foreign colleagues. Twenty years later, Simon, who had since achieved a high level of skill in martial arts, also resorted to this fascinating move when traveling in Morocco.

Marrakech, the third largest city in Morocco, means "the land of Gods" in Berber. This divine land is home to Jemaa el-Fnaa, the largest souk in the world. UNESCO had named the commercial market an Intangible Cultural Heritage site. Jemaa el-Fnaa was a must-see for tourists. A snake charmer held court at the most exciting and must-see booth in the market. It was said that the desert area of Morocco was rich in cobras, and snake-charming was actually a craft that has been passed down from generation to generation. As soon as the snake charmer plays the rhaita, a cobra in front of him will dance to the music. The tourists become fascinated, feeling a chill descend their spines. When Simon and his friends travelled to Morocco, they also went to the snake charmer's booth in Marrakech's Jemaa el-Fnaa. He asked for a small snake from the owner of the stall and performed a snake-shape hand. It was winter when he was in Morocco, and the weather was cold. Though he wore a thick overcoat and scarf, his moves were swift and agile.

The tour guide was a native Moroccan with a strong physique. He was almost 10 years younger than Simon. After watching Simon's performance, he said excitedly that he practiced boxing and would like to compete with him. Perhaps, all martial arts practitioners have a combative little boy inside. Simon immediately agreed. They stood facing each other. Simon darted like lightning, and deployed the snake-shape hand technique. Before the tour guide had time to react, the contest was over. Within the very few seconds, Simon made five moves: deciding where to hit; concentrating on the attack point; making use of the inner strength; making use of the martial arts skills; being adaptive to different situations. The mind, eyes, hands, footsteps and moves all need to work at fairly high speed in order to succeed. The rapidity arose from what Master Kwan called Chi: what can be loosely thought of as a sort of internal strength. The tour guide never managed to understand Simon's martial arts theory but admired his skills and asked to take a photo with him.

Simon not only "fights well", but also "runs well". In 2019, he inspected his new factory in Porto, Portugal. After work, local colleagues arranged him to visit a century-old winery. In order to maintain the right temperature for fermentation, the cellar was not very brightly lit. A tour group walked slowly in the cellar. The large oak barrels for making wine were placed high on both sides. A sandy lane cut through, forming an aisle. Perhaps, because even the molecules in the air were mellow with the aroma of wine, Simon suddenly grew excited, galloping across the coarse sandy ground of the wine cellar. Hearing the rustling sound, a colleague ahead of him turned around and shot the video of his dash. After a rough calculation, it was found that the speed was almost the same as the one he kept and broke the record in the middle school sports meet 45 years ago. His explosive power was amazing. Simon stopped to watch the video, feeling proud for having retained the speed of his youth.

When his companions uttered their amazement at his speed, Simon replied, "All actions originate from stillness".

He holds with a dialectical principle and was showing the surprising connection between stillness and speed.

The stillness is, for Simon, an important beginning for meditation.

In the Chinese classic book *Zhuangzi · Zaiyou*, When Huangdi, the Yellow Emperor, posed a question on philosophy to Guang Chengzi, a famous Taoist, he replied, "Ignore what you see or hear. Focus on meditation for self-cultivation."

Meditation creates a smooth flow of Chi, essentially the energy of life itself. Simon, as other skilled practitioners do, understands that meditation is not bound by time or place. Whenever he has the chance, no matter in the car, on the plane, or in the office, Simon can practice regulating his Chi — that is, meditating. It's just like what Simon said, "If you want to achieve a goal, you must seize every opportunity and try every possibility. Never give yourself any excuses."

Sometimes, this dedication can lead to unusual circumstance, such as losing his luggage while practicing meditation at the airport in the United States.

Once on the Greyhound bus from Maine to New York, he meditated for a full eight hours. The passengers on the bus looked at him curiously, not able to figure out what this Chinese man was doing. Simon remained stationary, not relinquishing his position even when the bus hit a rough patch of the road. While with Simon in

Yunnan, his friends recalled the time when Simon spotted a huge boulder under an old tree. Immediately, Simon climbed up its facing and meditated for a few minutes. When asked how he'd felt, Simon replied, "It's like sitting in the depth of silence and seeing a shining stream of light flashing by my heart."

Simon added, "Meditation, together with Ba-gua walking, is a perfect combination of action and stillness."

For over 20 years, he has each day practiced walking according to the Ba-gua method. The founder of Ba-gua Zhang, Master Dong Haichuan said: "Ba-gua walking is of utmost importance and extremely fundamental in the practice of Ba-gua Zhang." The footwork of Ba-gua walking is unique and ingenious. It's called the Mud-wading step. The feet are lifted, then land solidly as if gripping the ground. The practitioner of Ba-gua walking may feel a kind of Chi: an unceasing flow of energy: thus, with each step, feel his internal strength growing. In his well-known novel, *Demi-Gods and Semi-Devils*, Louis Cha described "Lingbo Weibu" or the "sliding over the water" technique as evoking a similar feeling.

With years of practice, Simon remained more agile than many young people. In his fifties, he could still do cartwheels. At the age of sixty, he went to Sri Lanka and did a handstand on the beach of the Indian Ocean. Without any support, he finished the whole process easily and rhythmically. He was very proud of his ability to see the world upside down, and jokingly said that he had a green old age or, in other words, an old age where he remains in the springtime of his life.

His physique has not changed much in the past 30 years. In 2020, with the outbreak of Covid-19 in Hong Kong, Simon stayed at home for a period of time. Thus, he had much time to sort through his old things. He found a Bossini iron gray Polo T-shirt that he bought over 30 years ago. He tried the apparel on, and it turned out that the T-shirt still fitted him well. He has remained in the same physical condition over so many years, but his mindset has changed greatly. When starting to practice martial arts, he had wanted to compete with others. However, the longer he practiced, the more he realized that the essence of practicing martial arts is to cultivate an inner virtue that cannot weigh merely in terms of wins and losses.

In 2018, Simon attended a gathering. One guest, a master of Tai Chi, gave a demonstration for the audience. Knowing that Simon also practiced martial arts, the master invited him to the stage to perform. At first, Simon refused but at last, stood on centre stage. The master asked Simon to grab his hands. Simon, who had practiced Ba-gua Zhang for many years, immediately realized the invitation had a

❖ *Ba-gua Walking is an essential practice for Simon every day.*

❖ *The harmonious combination of action and stillness*

purpose. The master wanted to compete with him in a contest to test their internal strength. If the master let go of Simon's hands, Simon needed to let go of the master's immediately without exerting even a slight hesitation. It's actually a form of competition. Although accepting the invitation, Simon never used his internal strength. The master was a bit disappointed and performed a set of Tai Chi alone.

Afterwards, Simon praised the guest for his professional Tai Chi skills and a strong foundation in internal strength. According to Simon, "The essence of Tai Chi lies in two words, motif and relaxation. In order to attain that dialectic, the master of the art must achieve a measured fluidity in his gestures and footwork. For this exercise, the diaphragm becomes an important vehicle. Only through deep breathing can one attain a unity between mind and body. I knew then I could learn by just attentively watching the performance of the Tai Chi master — and I was grateful." As to why he was unwilling to engage in a competition over internal strength, a source for Simon's hesitation can be traced back to an incident occurring some 20 years ago.

When he started to practice martial arts, Simon often met with fellow practitioners and sought all the opportunities to learn from his seniors. At that time, he was also combative, wanting to try out his own skills. At one gathering, when a senior who had been practicing martial arts for a long time wanted to find someone to demonstrate pushing-hands, Simon volunteered. Simon was straightforward and didn't think much. He went up, raising his hands to make use of his internal strength, pushing his senior back a few steps in full view of everyone. It was very embarrassing. Seeing the long face of this participant, Simon realized that his competitive zeal had led him to be disrespectful. Simon apologized profusely for his error.

Simon said, "Being reckless and competitive means an insufficient cultivation of martial virtue. Diamond cuts diamond. One needs to learn to appreciate those who are more excellent than he or she is. Modesty and open-mindedness are key-virtues to the practice of the martial arts. China's martial arts culture also has profound value, and each school has its own merits. What martial arts practitioners should prioritize is not the skills but the cultivation of inner virtue, a bottom line that should be defended!"

Simon was not only referring to martial arts training. The meaning went well beyond a simple practice. He added, "As human beings, we are learning all through our lives. We learn from mistakes, from experiences, from ourselves as well as from others."

Learning is indeed an essential part of self-cultivation. In Cantonese, people who have no formal training in management and start from the grassroots are called "Hung Fu Zai", which literally means red pants guys. Those who have formally learned the principles of effective management at universities are called academics. Simon, who calls himself "Hung Fu Zai", is highly skilled in the art of learning. Many universities nowadays are offering classes on management theories for entrepreneurs, but in the past, there were no such classes for Simon to learn. He said earnestly, "Even if there were, I probably wouldn't be able to afford the education. My method is to go to the society to learn more from experienced entrepreneurs, making friends with these masters, so as to find inspiration and thereby, to learn the way of management."

He is a great believer in the principle "experience is the mother of wisdom."

Young Law, the boss of Laws Textile Industrial Ltd., is known for being both sophisticated and a difficult individual. Once at a dinner gathering, he remarked, "Businessman should read bankruptcy law."

Young Law had no special intention, but Simon, who was also in this dinner, was inspired and impressed. Being mindful to issues that are easily ignored and staying alerted to risks that might occur are two critical points in the skill and art of

❖ *Simon (left) with Young Law*

management. However, Young Law identified both problems and solutions with a single sentence. This was exactly what Simon needed at that time when his business was growing in size.

It was not easy to get along with Young Law. Simon's key was sincerity. Although he has his own business and can be regarded as sort of a boss, he can swallow his pride.

He said, "I helped whenever I could, it didn't matter. Since I followed him every day, he could see if I was sincere! What's most unacceptable is that wanting to learn but refusing to make the effort. No pain, no gain. All masters will do something to test students. To be a student, I must follow a student's rules, but not act as a boss. I must, above all, be humble and sincere in my efforts.

"Attitude is of the first and foremost importance. Humble yourself and never mind momentary losses". Simon said earnestly.

It was often the case that Simon, while having a meeting in his company, was called by Young Law, who wanted to treat Simon to a lunch. However, after the grand meal, Simon paid the bill. "I was actually learning from him without paying the 'tuition fee', so it's quite reasonable that I pay the meal," Simon said. "Imagine that you want to learn martial arts, and you imitate some techniques in front of your master every day. Ultimately, the master will teach you his most treasured skills. If you are really determined to learn, it's enough to watch the master practicing for just a couple of times," Simon said.

Here is another dialectic Simon has discovered and practiced during learning: to lose can sometimes mean to gain.

"Restaurants, hotel lobbies, and trains all could serve as classrooms. Wherever Young Law went, I went with him. He would not hastily tell you his story, but as time went on, would definitely reveal a valuable bit of wisdom. You were learning enterprise management from an experienced businessman who had gained from both his successes and failures. So, he didn't easily tell you what exactly he had experienced, but once he did, you benefited a lot." Talking about the learning experience, Simon had his own feelings. "The most important thing I ever learnt from Young Law is that business management is actually human management. For this, he had his personal experience and methods, and I called him master of human nature management," laughed Simon.

"I've got opportunities to observe how he inspected his business, dealt with the government, customs, and other organs of the society. The quota he obtained determined how many raw materials he could import from abroad. These materials were initially processed in Hong Kong, and then they were taken to the Mainland for further processing until they became end products for sale. He needed to obtain further quota which would determine how many of these end products were allowed to be exported. Once you changed the location of processing, the profit would be very different. That is to say, when you rearrange your production locations and supply chains, in a legal way, of course, the profit you make will be very different. He can always think out of box," Simon said. "We are living in a big world. Don't involve yourself in what you don't know. If you want to join in any endeavour, you must be willing to learn from a master."

Suen Siu Wing recalled his older brothers' eagerness to learn and the singular approach Simon took.

"I remember when we were children living in the countryside, my brother was working hard on English. In our surroundings, no one studied English. People thought that if you learnt English, your mind must be impure. But my brother was unique. He had his own thought and vision. He knew what he was doing and where he was heading. As he didn't have the chance to receive school education, he asked dad to send him books, secretly of course; then, on his own, studied the language." At the time when many people simply had not a clue, Simon knew nearly two thousand English words.

In December 1977, Beijing TV, now China Central Television, was broadcasting an English-learning program entitled "TV Lectures on English". Since then, many have begun English, and more and more developed a passion for its study. Living in the countryside, however, Simon had no television at home. As soon as he knew the supporting textbooks and tapes for this program were available, he found someone who worked in the city to help for getting one set. That next weekend, a set of books, bound tightly on the seat of a bicycle, arrived at Simon's home after undergoing a two-hour journey. For Simon, these were not merely books. They contained his hopes and expectations for future

Published by People's Education Press, this set of textbooks contained, in total, four volumes. When one of them was discovered one day by surprise, Simon took it as a treasure. Although this book was a worn, its page yellowed from years of aging, his notes still clearly lined the volume. On that day, it happened that a friend came to visit. Like a child, Simon was excited to unfold the book and began to read the English sentences aloud, recalling enthusiastically the old days when he was a

❖ *The four English books that carried Simon's youth and dream*

diligent young student.

Simon tried to find the other three volumes to make the textbooks complete. When he finally got them, he was overjoyed for a whole day.

"When Simon moved to Hong Kong, he did not lose his passion for learning, and for knowledge. Because the education he received in the countryside was so limited, he grew appreciate the opportunity to learn and was sincere and modest in his desire to find a suitable teacher. This sincerity won him much support and recognition from teachers and friends across many different fields." Suen Siu Wing greatly admires his elder brother's respect for learning and thought the singular virtue was worth emulating.

Eric Suen was also impressed by his father's attitude towards learning. "Dad was not only doing business with others but also learning business management from others. For example, inspired by the business partner Esprit, he introduced the

'afternoon tea break' to the company, and also encouraged the management team to have 'coffee time'. No relax, no concentration."

"He is reading and studying all the time, and I really admire him for these traits. That may help to explain whoever dad meets, he always had a lot to talk and the topics can be broad and profound."

For Simon, learning alone is not enough. One must also be able to be enlightened.

"Fundamentally there is neither Bodhi-tree, nor stand of a mirror bright; since all is void from the beginning, where can the dust alight?" These lines show the importance of being enlightened.

"With these lines, Hui Neng had been able to change from being a simple farmer to be a most respected monk in the Chinese Buddhist history. How did Hui Neng achieve this miraculous metamorphosis? Enlightenment! " Simon always mentioned this example.

But in order to achieve such a transformation, Simon noted, "You should learn from others' merits. Their negative traits should serve as a warning. Enlightenment can help you think critically and enable you to see the difference and to discover the crucial lessons that allow you to become successful. In English, we say that 'There are a thousand Hamlets in a thousand people's eyes'; in Chinese, we can also say that there are a thousand understandings of *Romance of the Three Kingdoms* in a thousand people's eyes. Each person is a unique individual with the wisdom to achieve his or her singular path.

"Past experiences can sharpen your understanding but only if you reflect on that history, considering all the correct and incorrect decision made along the way. Only then can you blaze your own trail. Practicing martial arts and running a business may seem two distinct fields, but success in one field can inspire achievements in the other. It is important to see all matters as one in order to find one's own path towards success." Simon is good at using plain language and concrete examples to drive home his ideas.

Simon continued, "Chinese philosophy is inseparable from the interplay of the yin and the yang, the dynamic and the static, the hard and the soft, the tangible and the intangible. This is also true for practicing martial arts and running businesses, which is like waging a battle. The competition between rivals is no less than a psychological war that demands tactics and strategies. Any general knows it is important to remain calm no matter the situation but achieving calmness in a difficult situation is easier said than done, and a general must not only devise a single act but recognize that each act has an overall impact. So, sound judgment is a necessity in both the martial arts and in the world of business."

Simon has acted on that principle. It informs his singular attitudes, unique values and ultimately, his understanding of life. Simon believes that the perfect man ignores the ego and so, can see the world as it is. He has developed his unique outlooks on the world, life and value.

This is his way of looking at this world.

If one has a good conscience, he will have no obstacles in his mind. This is Simon's outlook on life.

Truth-seeking, fact-pursuing and value-persisting are what Simon deems the most significant thing to do. This is Simon's value system.

His three outlooks are concise and comprehensive, fully reflecting his exceptional wisdom gained from his meditation on the nature of existence. Chu Hok Ting, a Taoist priest, and friend for many years, praised Simon's outlook as logical, scientific, and philosophical.

Chu said: "Simon's self-cultivation is readily apparent. Firstly, no matter when you meet him, he is vigourous and in high spirits. Secondly, he never gives up. You never hear one word of discouragement or frustration from him. He always takes on a positive attitude. Many people may feel happy when everything goes right but become frustrated when meeting setbacks. This is because thcy haven't practiced, at least very well, the art of self-cultivation and therefore, are unable to attain success. Thirdly, Simon is never arrogant nor fearful. Based on my observation of him over so many years, I dare say his qualities as such have never changed."

Chu continued, "Some are likely to curry favour with those who hold power and have wide network but look down upon those who appear inferior. But Simon is not such an individual. This sense of humility embodies his maturity."

Born to a family of metaphysicians, Chu Hok Ting looks, despite his age, to be young and vigourous. He has taken on a Taoist hairstyle, wearing Buddhist beads and the clothes of a Confucian, with a powerful voice.

"Simon's wisdom, courage, and heroic spirit speak for his adequate self-cultivation, which is also a primary reason for his success."

Simon also prefers to use the word "self-cultivation" to describe the requirements for himself.

Simon said, "No one is perfect. We must all recognize the fact and overcome our shortcomings by using correct methods. This is 'self-cultivation'. Without self-cultivation, one can hardly have any achievement." Simon confessed that he

❖ *Simon and Taoist Master Chu Hok Ting (right)*

occasionally grows irritable but would reflect on self immediately, admitting that he had not cultivated the self sufficiently. It is adversity, not prosperity that brings forward one's self-cultivation.

In 2020, an unexpected outbreak of the Covid-19 pandemic disrupted our life. Simon, however, remarked that he felt a sense of tranquility and profundity. The pandemic earned Simon a lot of spare time. He practiced walking and meditation as usual, regardless of the change in the outside world. He gained the chance to read, attended online lessons on traditional Chinese culture on a weekly basis and began to practice Chinese calligraphy. Simon, used to coping with changes by changing, said that this was his unique attitude towards the pandemic.

Simon had never practiced calligraphy before. His first try was the time when he was enjoying his holidays at Mount Lu in central China. There he saw a folk artist who earned a living by composing acrostic poems for tourists. The rule was that, in a poem, a tourist's name must be creatively embedded. Simon was very interested in the skill and asked the price. The artist replied: "Two hundred RMB for one poem. I assure that you will be satisfied". Simon laughed: "I give you two hundred RMB to buy a piece of paper, but I will compose the poem and write it down by myself." He then wrote a line, comprising ten Chinese characters, with a brush-pen. The two characters "Siu" and "Man", his name, were embedded in a poem about a beautiful natural scene and a man's virtue. Simon's brushstrokes were strong and vigourous, reflecting his study of Chinese traditional culture and his own character. Passersby stopped to watch Simon writing, admiring his skill greatly.

Having practiced calligraphy for some time, Simon began to draw impromptu pictures right next to the characters he wrote. With just a few brushstrokes, he depicted a countryside scene with hills, water, boat, bridge, house, sky and eagle.

After this first attempt, Simon's passion for painting grew in intensity. Once he painted until four o'clock in the morning.

His palette was quite simple: just a small brush-pen, which he used for both calligraphy and painting. The simpler a tool, the more pleasure one can derive. He would apply colours on a tissue first. Upon completing a work, the tissue paper often looked like a rainbow. Some artist-friends praised his work as unpretentious while effusing an unforced style consistent with Chinese traditional paintings. Simon said that he had only bare any formal training. His paintings reflected his own playfulness. "I paint just like a child, without any thought as to what was good or bad — just for my own pleasure."

Although he had not been formally trained in painting, Simon did have a great deal of experience as a collector, and no doubt, his unforced archaic style grew out of this exposure. During his childhood in fact, Simon had avoided lessons in painting, and it was only the pandemic that enabled Simon to harvest his natural skill.

For Simon, painting and calligraphy became yet another way towards self-cultivation.

Simon said: "Your attitude towards setbacks determines how you will lead your life. The more powerful you are in the face of setbacks, the farther you will go. Reading, calligraphy, martial arts, and painting all represent approaches to self-cultivation. If you persist, they will cleanse your mind, boost your spirit, and strengthen your immune system."

He often shared with others his unique way to address the changed situation brought about by the pandemic. The Chinese classic, *The Great Learning*, notes: "When you know where to stop, you have stability. When you have stability, you can be tranquil. When you are tranquil, you can be at ease. When you are at ease, you can be purposeful. When you can be firm in your efforts, you can attain your aims."

Fools view their environment as a constraint. The wise see their environment as a resource to be taken advantage of. Being foolish or wise all depends on your state of mind. Through years of self-cultivation, Simon has changed from an aggressive and anxious individual to a leader who is composed and at peace. His wisdom becomes more necessary and apparent in times of adversity.

The book *Doctrine of the Mean* says that "Sincerity can automatically lead to success". It means that one should cultivate this fundamental trait. As long as you are sincere, you will definitely get a good result, just like the saying in the same book that goes: "Those who are sincere will definitely be successful."

And so did Simon.

❖ *Simon's Painting*

❖ *Simon (third from right), Simon's father Suen Shing (fourth from right), and Simon's siblings*

CHAPTER

# VII

## MORALITY

/

## MANAGING FAMILY

# A Harmonious Family Brings Prosperity

Discovery Bay is a low-density residential area located on Lantau Island in the northeastern region of Hong Kong. Surrounded by sea on three sides and not far from Hong Kong's central business district, it looks like a secluded paradise. Every year on the ninth day of the first lunar month, Simon would ask his driver to take him to the Central Ferry Pier No.3, where he would take a ferry to Discovery Bay to celebrate his father's birthday and the Chinese New Year as well. Many of his siblings would also be there. For years, as the oldest brother, Simon would organize the gatherings.

The grand lunch was unveiled in a Cantonese restaurant near Suen Shing's home. Two big round tables were placed in the lunchroom. The senior generations sat at the inner table, while the young people sat at the one near the door. The dishes were served soon after. Following the Chinese custom, Suen Shing sat on the most respected seat at the innermost table. Even though he had achieved old age, Simon's father still cared very much about his appearance. His hair was well combed; he wore a blue down jacket and a fashionable red, green, and brown scarf from a famous brand that Simon had bought on a business trip to Italy. Suen Shing held in his hand New Years' red packets. He gave everybody a packet. Afterwards, there were still some red packets left in his hand, so he asked: "Does everyone have a red packet? What about waiters?"

When he was young, he was known for his generosity. The trait has not dimmed over time. Simon's filial piety and respect for his father can be clearly observed on such occasions. When his father was speaking, Simon listens attentively. Even back to the early time when he was poor, he treated his father with respect. Later, when he made his fortune and provide his father financial aid to weather difficulties, Simon still showed consideration and regard. Now although his reputation has grown as his business has expanded, Simon's humility and piety before his father has remained unchanged.

Suen Shing married again after relocating to Hong Kong. The new wife, also Simon's stepmother, once wrote Simon a letter, at that time she was suffering from cancer and that Suen Shing's company had been long on the brink of bankruptcy. Simon immediately took on the medical expense, although he also needed the money for his own business, which was just beginning. Simon never spoke of his difficulties. In the letter, his stepmother expressed thanks for him, and also showed her sympathy for his difficult childhood and his lack of parents' love.

"When I was a child, my father was not living with me. My early memory of him was quite vague. I hardly knew him until I began writing to him. But his letters taught me a lot, especially how to behave. He rarely sent money home except for once when he bought me a pair of sneakers. I was just starting to practice track and field and saw that the other kids were wearing professional shoes. That fascinated me, so I wrote my father asking for a pair. I can still remember how happy I was to receive the gift." His friend Chen Manlin joked that Simon's shoes were made of genuine leather and did not reek of a bad odor whereas his were made of artificial leather, which could grow smelly.

"What my stepmother related was true. But I never held any grudge against my father. Rather, I have always been grateful to him. My gratefulness has nothing to do with money or daily necessities. I am indebted to him for the opportunities that he gave me, which allowed me to shape my destiny."

The first opportunity his father gave him was the year 1980 when Simon arrived in Hong Kong, and the "snakehead" called his father asking for money. His father was able, with considerable pain, to gain the two thousand Hong Kong dollars. With the money at hand, his father rushed to the bus station in Tuen Mun and set Simon free.

"He had been trying to persuade me not to come to Hong Kong, but when I really did reach the island, he tried his best to raise money and pay the ransom." Simon was fully aware what the two thousand Hong Kong dollars meant to his father, whose company was teetering on the edge of bankruptcy. Back to the year 1980, no one was rich. It was not easy to borrow money, especially such a large sum in a really short time

"If my father didn't raise the needed ransom, I would never have the chance to gain a temporary Hong Kong ID and would never have the chance to set up my own business in Hong Kong. All my later achievements are indebted to this significant 'if'." It may be commonplace to believe that a father should naturally do so for his son, but Simon never took the action for granted.

"After my father ransomed me, he arranged me to live with his friend. Before he left, he gave me a few words: 'either sell fruits in Mong Kok; or you will be a hooligan.'" His father knew that Simon back in the Mainland was skilled at provoking trouble. Yet the next day, his father introduced him to The Mira Hong

Kong hotel, as a waiter.

"He had already tried his best. I stayed in his friend's home for two days; then I left, knowing from that moment on, I had to rely on myself."

After serving as a waiter for half a year at The Mira Hong Kong, Simon hunted for a new job as a courier in OCS. Then he became a salesman at Fair Label. Later on, he set up his own business, becoming a broker. In 1983, Simon planned to go to the Pearl River Delta region, an area of the Mainland neighbouring Hong Kong. His father helped Simon to gain connections with businesses in Dongguan, Guangdong Province, which significantly opened up a new prospect for his business.

"My father had worked in the Mainland and had a network there. As soon as he learnt that I wanted to go back to the Mainland to set up my own factory, he took me to Dongguan himself."

With his understanding of the industry, as well as his experience and passion, Simon was successful in forming a partnership with the Dongguan government. However, he was always grateful to his father for introducing him to his network of contacts. "If my father didn't pull strings for me, I would probably not be able to have that opportunity." This is what Simon always calls the second chance his father gave him.

The third chance given to him by his father was an economy class air ticket to the US.

"After I had with the Dongguan government set up a joint venture, my father asked me one day: 'Do you want to go to the US to see their label companies?'"

"Of course, I was eager to go. However, my stay in Hong Kong was too short to entitle me to a permanent citizenship, so it was difficult to apply for a US visa." Luckily, Kincheng Bank, now Bank of China (Hong Kong) Ltd., had been willing to serve as Simon's guarantor. As a result, Simon was able to gain his US visa.

With the visa at hand, Simon went to the US with his father where he became acquainted with Rollin Sontag, the boss of ASL, a successful US label company.

One's vision determines how far he can go. Only when you see with own eyes real excellence can you begin to know how to define it. "The air ticket my father gave was very significant." This trip profoundly changed Simon's understanding of the label industry and inspired him to have long-term and innovative vision. He decided to aim high — to make a global business and become No.1 in the field.

Suen Shing also remembered this trip, saying: "We also made a trip to the UK together later on. He wanted to buy the state-of-the-art looms."

Although Suen Shing was over 90 years old, he could still remember the exact date, month and year when talking about the old days.

"I arrived for the first time in Hong Kong on the $15^{th}$ of August in 1935. It was my adoptive mother who took me there." Suen Shing made a living in Hong Kong when he was young, but as soon as the Korean War broke out in 1950, he went back to his hometown Chang'an in the Mainland, hoping to join the army and serve the country. However, by accident, he became a teacher, and later on was even promoted as the school principal. In 1962, Suen Shing returned to Hong Kong and ran a trade company. He managed it so smoothly that he was able rent two rooms for an office in the Dong Ying Building. "Back in the 1970s, the two most splendid buildings in the entire Kowloon were the Star House and the Dong Ying Building. Most of their rooms were occupied by airlines, foreign companies and private doctors. Very few trade companies could rent a room there, but I was the exception."

Suen Shing was in the apparel export business. After he received an order, he outsourced the production work to an agent and made a profit from the difference in price. His business philosophy was that "those who can help others achieve their goals will win all." Simon's father understood, then, when his business partners profited, he could build a profitable business. He carried this principle to the next level, letting his business partners know the initial profit. His transparency created trust. Soon, his reputation for fairness got out, and the numbers of orders he received increased. Even though his profit margins were relatively small, his company grew in size and reputation. During a peak period, his company even attracted orders from Lian Fang, the biggest export business then.

"My father is good at communicating with others and is fully able to expand his business. His success lies in his 'aggression'. When I began to do business myself, I felt that I should not only be aggressive but also try to minimize risks when possible or be defensive. The balance of the two can contribute to the long-term development of a business." Simon acknowledged the influence of his father — in particular when it came to running business.

"As a young man, Simon was handsome. My friends told me that my son didn't need to make any effort to get an order. Every time he went to a foreign company, the agents, young ladies, were eager to give him orders." Suen Shing continued, "Of course, this is a joke. He has his own business philosophy and didn't follow others' footsteps while sticking to a strict ethical code. Moral integrity matters a lot when dealing with others, especially with your customers. Simon maintains good terms with his customers, so his relationship with each customer can last a long while."

"Simon is smart, wise and hardworking. He experienced ups and downs in his career, but for him, the losses do not matter; what really matters is whether you have the audacity to suffer losses." Suen Shing read books on Marxism-Leninism and Confucianism when he was young. His ideas on running business are profound.

"When Simon was young, he easily lost his temper, even pounding desks. I knew that he was under pressure, and sometimes the pressure was difficult to manage". Suen Shing understood his son's anxieties and, after an outburst, would note that, "In fact, he always shows his respect for people, especially those who are knowledgeable and hard-working. If you are really outstanding, he respects you so much."

No one knows a man better than his father. At his home in Discovery Bay, Suen Shing placed a photo taken with his son in the most prominent place in his living room. The silver-coloured picture frame is thin and delicate, and in the photo, the father stood shoulder to shoulder with his son, who wore an honorary doctoral regalia. They laughed happily.

"Simon shows great respect for his parents, especially for me," Suen Shing raised his voice. As a son, Simon abides by the principle of filial piety. That's for sure.

Although he only earned several hundred Hong Kong dollars when he just arrived in Hong Kong, Simon would take his father for yum cha, and before they parted, he never failed to give his father one hundred dollars for taking taxi. When his career took shape, he paid office rent for his father every month, and also left some money. When he finally had his credit card, he assigned his father as a secondary cardholder. When his label business became so huge that his photo appeared on a magazine's cover, his father looked at the cover and couldn't help but shed tears. After all, when a son accomplishes a great feat, the happiest person should be the father. When he became dedicated to public welfare undertakings and often participated in art exhibitions; Simon also invited his father to be present. On these occasions, Suen Shing always took a small camera with him to capture pictures of his son's accomplishments.

Looking at the silver picture frame, Suen Shing said: "Simon also took good care of his sisters and brothers. He has never hesitated to lend them a hand whenever he thought necessary." Shortly after the New Year gathering, his younger sister called him, saying that her husband recovered very well after the rehabilitation care by the doctor who Simon had introduced.

Suen Siu Wing, Simon's younger brother who grew up with Simon in the Shangsha village, described Simon as a very filial and responsible person.

"In 1962 when our father went to Hong Kong, Simon was just 5 years old. Our father had not been with us for over ten years. To be honest, we didn't have any particular affection for him. Only after 16 or 17 years did we meet our father again. Many people didn't believe that we could maintain a good relationship with our father under such circumstances. However, Simon did a brilliant job in this regard. In fact, before we came to Hong Kong and met our father, we had thought that he was living in a very good environment, but when actually arrived in the city, we found surprisingly that his living environment was very poor. After Simon settled down there, he covered many of our father's expenditures. After our father retired, he often got sick and needed to see a doctor. It was Simon who took care of all the expenditures. In a similar situation, many other families would have broken apart. But my brother bears a traditional attitude towards the family. He is willing to take responsibilities without any complaint. That is a rare attribute. He is the oldest sibling, and never fails to show his responsibilities as a big brother. Most of all, he is always bringing us together, making sure we are all close. I believe especially in modern times, such care and concern towards the family are not commonplace. He is an extraordinary brother and son."

Several years ago, Suen Shing was caught between two doors of an elevator, suffering from hip fracture. He needed a wheelchair and became too depressed to go out. Full of vigour and vitality in his youth, he couldn't accept the fact that he had to rely on a wheelchair. Simon empathized with his father's feelings and pushed him on the wheelchair for a walk or for yum cha in the city centre. Simon believed that his father might have a better mood after experiencing the hustle and bustle of the downtown. Shortly after, Suen Shing became cheerful again, so Simon encouraged him to do some very basic exercises to maintain his health. As his father gradually recovered, he encouraged his mother to combat illness by using the same method.

At the age of 19, Simon met his biological mother for the first time.

"When I graduated high school, mom was anxious over my job prospects. She was afraid I couldn't find a job and urged me to study Chinese medicine."

But the profession required its practitioners to have a peaceful and tranquil mind. After two months burning wood to boil medicinal soups, he was convinced that he was not made for that occupation. Simon quitted. However, he did not forget the classic books on Chinese medicine he had studied during the period. Sometime later, the pursuit proved helpful in his health management. At times, Simon even taught his friends how to practice what's called Wu-xing or the Five Phases, a traditional treatment and philosophy centring on physical and spiritual balance that is traditionally used to maintain good health.

"I am indebted to my mom for her decision".

At the end of 2017, Simon's mother suffered a stroke and couldn't take care of herself. As a person with a strong character, she became greatly frustrated and easily lost her temper, often quarrelling with her personal nurse. When Simon visited his mother, he could see the decline in her spirits and became dismayed.

Simon asked his younger brother in Dongguan to find another nurse for their mother and arrange for her to receive Chinese acupuncture treatment. He empathized with the pain that his mother was experiencing. When she was young, her love had not brought her happiness, and now when she was old, poor health resulted in her feeling ache whenever she moved. Simon tried his best to make her happy and decided to throw her a grand party for her 80th birthday. In the middle of March, as soon as he arrived in the town from Beijing for a meeting, he began to prepare for the celebration. He also invited his colleagues in Dongguan to join, because he believed that the elderly preferred lively atmospheres. Unsurprisingly, Simon's mother was very happy. Taking Simon's hands, she talked to him lovingly and glowed with happiness. After the grand dinner, Simon's son and two daughters saw off their grandma, who enjoyed talking with the three grandchildren. Simon's children were all present, except the second daughter who was in the UK.

Lesley, Simon's youngest daughter, confessed: "When I was a kid, I couldn't understand why my father bore all the responsibilities for his parents who didn't even accompany him from childhood to adulthood. He has been a financial pillar, and he spent a lot of time, taking care of them."

Among Simon's children, Lesley looks the most like him. If you put their childhood photos together, you will immediately notice their resemblance. "But daddy is never calculating and has been very filial to my grandma and grandpa. His conduct has provided a model for me." For the Chinese, filial piety is the most important virtue. "My daddy seldom instructed me on what I should do, but he simply set an example for me to see and to feel."

"At a very young age, we began to have meals with adults, no matter the occasion. Dad never said that kids should sit aside, always sat together." Chloe,

❖ *Simon and his mother*

Simon's oldest daughter, recalled. "Maybe my parents did so because they were busy, yet we, as their children, could see clearly what kind of people they were, and how they dealt with others. Through their example, we came to understand basic social etiquette." She added, "Daddy wouldn't criticize us even if we interrupted their conversation. He never thought that kids couldn't talk at meals. He respected us a lot."

Simon is perceptive. As a father, he listened carefully to his children's words and was able to recognize the uniqueness in each of them at their young ages.

"When I was in my junior high school, daddy told me that I would be destined to do marketing jobs."

Lesley is tall and fashionable. After graduating from the university, she worked for an international advertising agency. "At first, I didn't understand daddy's judgment and didn't want to be labelled as a person who could only do marketing. Only in recent years did I realize that what he meant by marketing was not all about advertisement, but about his judgement that intuitively, I understood how to sell products." Lesley admired her father's insight. "I began to study abroad at a very young age. Daddy was not that kind of person who called me frequently. Although he was not around me, he knew everything about me, my strengths and which careers and industries they matched. All his judgements were accurate. Amazing! "

Speaking of telephone calls, Chloe was reminded of an incident.

"Duing my junior high school years, communication was not as convenient as it is today. If I wanted to make a phone call from the US to Hong Kong where my parents lived, I needed to buy first a telephone card, then, to visit the school telephone switchboard and call them. If the timing was not good, I mean, no one was at home: it was very difficult to reach them. But it was even more challenging for them to call me because the school campus was so large that my parents most likely couldn't find my whereabouts: whether I was in classroom, at the canteen, or in the dormitory."

"If my memory doesn't fail me, it was one evening of May. But in Hong Kong, it was in the morning. I was dining in a school canteen, when a teacher suddenly came to me, saying: 'Go to the telephone switchboard immediately, your father is calling you.' I ran all the way to the switchboard and picked up the phone. The first sentence my father said to me was: 'It was Mother's Day. Why didn't you call your mom?'"

Chloe said that her father has never shouted at her except this one time, so she remembered it clearly.

"In educating his children, my father has a full picture. For daily trivialities, you can have much freedom, but for principles and values, especially moral integrity, you definitely must do what is proper. For example, when he decided to send me abroad for education, the most important matter for him to do was to find a good school for me, because a good school teaches students by following strict regulations. He never cared about issues like when I went to school, what time I went to dormitory, and what kind of friends I made. This was the freedom he gave me. However, he didn't allow me to forget to call mom on Mother's Day. I needed to remember that mom was thinking of me every day in Hong Kong. In his eyes, honouring mom with a phone call was a moral duty. Back then, it was not easy to make an international call to my school and reach me, but he insisted on doing so. Therefore, I describe him as a father who balances freedom and strictness."

Simon agreed with what his daughter said about him and kept it in heart for years.

His younger daughter Nicole said she inherited a high level of moral integrity from her father. When she was a kid, if she saw her parents run the red light, she would get very angry and criticize them. She also learned business ethics from her father when working at his company.

Nicole said, "My father regards a clear conscience as a prerequisite to becoming a capable businessperson."

Eric, Simon's youngest son, admired deeply his father's integrity.

"Father follows moral codes in everything he does, whether in doing business or in his personal dealings. I have learned from him that making money is important but not the most important attribute. A gentleman makes money in a proper way. Money made from society should be used for society. His business philosophy has a decisive influence on my values."

"Dad has been teaching us not to forget where we come from. We should be grateful for what we have and give back to society whenever we can," her father's sense of altruism impacted her greatly.

She accompanied her father to many charity events. "Dad believes that society provides opportunities for him to thrive and it's his responsibility to give back to the society through public benefit and charity activities."

According to the annotations to *The Rites of Zhou*, "De Xing is the matching of the inner and the outer: what exists within one's heart is De (character); when put

❖ *Simon brought Lesley and Eric back to the Shangsha Village and took a photo in front of the house he used to live during his childhood.*

into practice, it is Xing (conduct)." Morality provides a compass providing the scope of one's deeds. Instead of lecturing his children about morality, Simon adheres to high moral standards and set examples by his own deeds in daily life.

Chloe recalled that Simon would often put himself in other's shoes. He would not just think about himself. Her father believed that what was good for others was also good for the self. Simon's philosophy was based on the principle of benevolence in the Chinese culture. Western culture values self while Eastern culture emphasizes the world. Fu Xuan of the Jin Dynasty wrote, "A benevolent person puts himself into others' shoes. Understanding his own needs, he will try to meet the same needs of the populace." The ancient Chinese sages uphold benevolence and consider others before they act. Simon's adherence to benevolence has strongly impacted his children.

Nicole recalled, "Whenever he visits other's house, dad would always bring a gift. Such basic etiquette shows your sense of ethics. Now I would do the same thing." Lesley agreed, "Dad was very generous." Her generosity is inherited from her father. When buying coffee at noon break, she would buy one for every colleague in her team.

"The most important lesson I learned from dad is that we should be willing to suffer loss, especially as a man," Eric added.

"Dad regarded morality education as extremely important." With her ponytail, Chloe still looked like a student in her white T-shirt and jeans. She managed the foundation under her father's name and founded a private art museum. She participated in many cultural activities and laughed that she had to wear heavy make-up so that would not be mistaken as a new young staff in the museum. Chloe was good at impromptu speech in Chinese and also spoke English very well, which was elegant and full of quick wit. It's hard to believe that she hadn't learned any English before she went to America. Her first year in America was extremely hard because the life and study there was totally different from what she experienced back in Hong Kong.

Chloe said, "Dad accompanied me to school on the first day. During my early days in America, he came to see me regularly. He carried along no luggage and flew in on the plane on Friday, and would spend two days in America with me and then fly back to Hong Kong. In the past when life was not as good as it is now for most people, few parents would go to great lengths to fly to America for their children's first day of school, let alone to attend the school's open days. However, my dad attended all the school activities. I could not imagine how busy he was but was really grateful that he could spare time from his tight schedule to support me."

"He kept encouraging me, asking me to study hard and not to give up."

After one year, Chloe caught up on her studies and obtained good grades.

"When my younger sisters and brother studied abroad in America, dad didn't come to see them regularly. It was not that he didn't care. He trusted me and believed I would take good care of them."

As the first child, Chloe had most time with her father alone. "On weekends, we would go to yum cha in the morning and then to the bookstore. Dad was frugal, but he would never refuse to buy a book. We bought a lot of books. Although mom felt we bought a bit more, but would not say a word. After the bookstore, we would go to the market and then cook dinner at home. Every weekend was spent like this. Actually, dad was very good at cooking. He could even master hard-to-cook dishes like pork bones porridge and steamed crab, super tasty!" Nicole also praised her father's skills in cooking fish. "Dad's fried prawns in soy sauce were even better than Mom's."

"I heard people say that dad was short-tempered when he was young, but I had never felt this way. He was very patient with us," Chloe recalled. Lesley agreed, "Dad never scolded me, let alone hit or beat me. He only spoke loudly to me once. One dinner, he blamed me for not eating the pumpkin in my bowl. I cried; after that, he said nothing." Lesley laughed.

"Dad was very reasonable and sensible. He would ask for our thoughts even if we're just kids. He respected us."

Eric is the only son in the family. He is tall and looks like his grandfather. "I also remember dad only scolded me once. He would not remember this. When I was a kid, I liked to eat bananas. Dad said it's not good to eat two bananas in one day, but I wanted to eat a second one so much. He said something which I didn't remember, but I was convinced. In the following two decades or so, I never ate two bananas in one day."

"Perhaps I was the only kid who saw dad's short temper," Nicole said.

Nicole exercised regularly and looked very fit. She was quite straightforward, "Dad and I have similar personalities. We stick to our own principles and don't make compromises easily. Usually we get along well, but when we disagree with each other on matters of significance, we'll have a fierce quarrel."

Nicole recalled that she didn't want to study abroad while her father insisted that

❖ *Chloe spent a lot of time with parents in her childhood.*

❖ *Lesley said Simon had been very patient with them since they were kids.*

she should. They quarreled with each other fiercely, and she refused to talk with him for several months.

"This is the only decision that he forced me to take in so many years, but I'm really grateful for that, and also felt very lucky, as his decision actually has changed my life."

Nicole is like her father. Both of them had a sweet tooth.

"When I recommended Krispy Kreme donut to dad for the first time in America, he refused to try the treat; when I bought some home, he couldn't resist, eating one after another. Every year in May and June, we would eat lychee. Dad and I really had trouble resisting sweet food. We just couldn't stop and keep eating until we had heat rash." Nicole recalled this sweet memory happily.

"Dining together is a huge event in our family. We'd go to yum cha every weekend and celebrate everyone's birthday by dining together. We ate together on Father's Day, Mother's Day, Mid-Autumn Festival and Winter Solstice Festival. This has been a family tradition." Chloe added, "Although dad is very busy, he would spare time to stay with us. This is very important to us. What is a family? A family should always eat together!

❖ *Simon kept the family tradition to dine together on important days.*

*The Book of Rites* says, "Etiquettes and customs begin with dining activities". Ways of life can be learned at the dining table. "Although we weren't born into a scholar's family, we're taught to be respectful to the elderly and would pour tea for our grandparents at the dining table. While dining with our big family, we also learned a lot from the stories of our relatives."

Simon's value of etiquette in dining has been passed down to Chloe. "My friends know how much I value rituals for dining. Entertaining guests to a meal is a huge matter. I will make careful plans about how to set the table and how to arrange the seats. Dad had his own way of teaching us, not just about dinning: also, for example, reading."

"Dad put his large bookshelf in my bedroom. You could find almost all types of books here: management, psychology, traditional Chinese medicine, Tsai Chih Chung's comic books, Zhang Xiaoxian's novels or even books about table manners." Chloe still remembers all these books. "Dad never asked us to read this or that book. Right after I began to learn reading, I started to browse through his collection."

"One's wisdom is embodied in their bookshelf." Chloe had a keen understanding of the importance of reading.

"When I was eleven or twelve, I began to read *How to Win Friends & Influence People*. The skull on the black book cover attracted my attention. Dad and I both liked reading Tsai Chih Chung's comic books about Chinese philosophy and history. In fact, Tsai's works served as my philosophy primer. But I was most interested in psychology." Psychology became Chloe's major in university.

"Dad never forced us to read. Instead, we saw he was an avid reader, and we learned from him by example." Chloe said, adding that going to bookstore and yum cha were their traditional family activities on weekends. Both events became a part of a long-cherished tradition.

Eric recalled as a kid going to the bookstore with his father where he developed a lifelong fascination with reading.

"Dad was such a huge influence to me. He often talked about the books he had read and naturally I grew interested in those books." After hearing his father talk about *Water Margin*, *Romance of the Three Kingdoms*, the two of China's four great classic novels, Eric, by then only a middle school student, bought all the four novels and read them one by one. "Few of my peers had read all the four classics. I was really proud of myself."

When he was in a good mood, Simon liked to recite Yang Shen's *Lin Jiang Xian*, a poem Eric also liked. After reading the novel *Romance of the Three Kingdoms*, Eric grasped an essence of the poem: "it conveys a heroic spirit." Eric wrote the whole poem down on a piece of paper and pasted the slip on the base of his table lamp. "The paper is still there." Eric's interest in reading gave a solid foundation for his language and cultural competence, which is why after years of studying abroad, Eric remains very skilled in Chinese.

Although Eric is no longer the boy following his father to the bookstore, his passion for books has remained unceasing. Whenever he finds a good book, he would buy an extra copy for his father. Recently Simon had been reading a volume on Artificial Intelligence that Eric had given him, knowing his father's interests in advanced technologies.

Nicole said she liked to read historical books and biographies. These books helped her to learn from the past and the famous historical figures also motivated her to become more self-disciplined.

Lesley was not particularly drawn to reading.

Chloe said of her father's evenhanded attitude, "Dad would never apply one set of standards to teach all his children. This could be very difficult for some parents, but not for mine. My parents would not compare one child with another, so we didn't feel a sense of competition. In terms of academic performance, dad expected more from Eric and me. He knew we could do better. If we could get 90, we would get criticized for only getting 80. Lesley was different from the rest of us. Her talent was not in study. Dad realized that a traditional exam-oriented education did not fit her and sent her to America much earlier than the rest of us."

"Dad's parenting style was based on positive enforcement."

Chloe was using a psychological term to describe Simon's parenting style. "He didn't supervise our homework every day. Instead, he gave us enough freedom as long as we're on the right track. If we did a good job, he would praise us," Chloe said. "We didn't worry about getting criticized by our parents if we failed in an exam. We wanted to do well and win their appraisal."

Eric also benefited from his father's use of positive reinforcement.

"Dad never forced me to recite the works from the *Three Hundred Tang Poems*. However, I really admired his ability to recite traditional poetry and began to practice myself. When I gave an impromptu recitation of the poems, dad was greatly impressed."

Chloe smiled when she talked about how each of them absorbed in their own way their parents' values and code of hard work. "Perhaps our own self-motivation assured our parents that we would use our freedom wisely. They trusted us to choose our own university, major and respected whatever we would like to do. However, whenever we needed him, he would give us a helping hand."

"I remember I was not very happy about my first job after graduation in America. My fellow colleague got promoted, but I didn't. Although I didn't quite like the job, I wished to get a promotion in order to prove my worth: then I would quit." When she shared with her father the plan, Chloe recalled, "He simply said, 'you're wasting your time because you didn't have any plans after the promotion, nor did this promotion have any positive impact on your long-term career development'". Simon's words were a wakeup call. Chloe decided to develop her career back in Hong Kong immediately.

Lesley also benefited from Simon's straightforward yet very sensible suggestions.

"Dad never asked me to work in his company. When I first started my jewelry brand, my business partner had another job, so I thought I should also have another one. As I majored in advertisement, I applied for many jobs in advertisement. However, after the interviews, I was not satisfied. Dad knew that I was looking for a job and asked me whether I would consider working in his company. I remember it was Father's Day. He seemed to have read my hesitation, telling me to just have a try, and could leave if I found a better choice. His words dispelled my anxiety and relieved my pressure. I felt like I was just working for the company, not for dad." Lesley added. "Actually, dad has never given me any pressure in so many years."

Nicole shared everything with her father, whether it's about work or personal relationship. Whenever she felt unhappy, she would talk with her father. After a year's work in Los Angeles, she hesitated about whether to return to Hong Kong. "Dad asked me two questions: first whether I liked my current job; second whether I could apply my strengths in doing the job. He reminded me that both passion and competence were essential in one's career development. He asked me to stop for a while and whenever I felt lost, to think carefully over the matter. He supported whatever course I set upon. This was really important to me."

After working in Hong Kong for several years, Nicole went to the UK for further study. On the day she left Hong Kong, Simon wrote a poem for her:

*Embarking on a journey far away from home,*
*To pursue a dream that inspires you all along.*
*Hard work will never fail you,*
*no matter how many hardships come along.*
*Long, long will be the road ahead of you.*

This poem expressed a father's heartfelt wishes to his daughter: I would miss you once you leave for the UK, but I admired your perseverance and ambition in pursuing your dream. If you kept working hard, you could make progress every day. No matter how many setbacks you had gone through, you should always keep a positive attitude. Even when you made some achievements, you should stay modest because success was a long way off.

On Nicole's birthday that year, this poem became a source of inspiration for the theme of her birthday party.

"For dad, family means a lot," Nicole said, "and I really appreciate that quality. This appreciation becomes stronger when I grow older. Life is like juggling balls. Dad knows very clear what is more important to him. He is good at balancing family and work. We didn't stay with him every day, but whenever we needed him, he was always there at our birthday parties, graduation ceremonies — whenever there was a significant moment, dad would always be present."

"Dad attended all of our graduation ceremonies, from high school to university," said Lesley. "His presence made me feel all the more each ceremony was a milestone in my life."

In 2019 when Nicole graduated from Imperial College London, the graduation ceremony was held in Royal Albert Hall, which had been a long tradition of the college. Set up in 1871, this red-brick building is the oldest music hall in London. The exterior of the hall is modelled on the colosseum, and the dominant colours of the interior design are red and gold. This solemn and magnificent hall is an expression of historical profundity.

Simon sat close to the stage. While he was watching the ceremony, the camera suddenly focused on Nicole who then appeared on the big screen. Simon captured this moment in a photograph and with the words: "the loud applause is for you, my dear. Today you have realized your dream, yet a new dream is coming. Keep your creativity and pursue your own career. From dad at the Imperial College London."

Eric too idolizes his father, learning important life lessons from this mentor:

❖ *Simon captured the shot of Nicole on the screen.*

"His love for and responsibility to our family influences me deeply. For me, family always comes first. I like to celebrate with my parents on my birthdays."

"If you look at all his experiences and achievements," Eric added, "he is a success by any measure." This might be a son's highest compliment to his father: to hold him as an idol.

Simon seldom talked about the difficulties of his childhood with his children. "Dad was very empathetic. He knew it was not reasonable to force us to live a hard life like he had suffered. We're two different generations." Chloe admired her father's open-mindedness.

"Dad never forced us to do anything." When he was a teenager, Eric learned Wing Chun, a traditional Southern Chinese style of Kung Fu. Some would think it's his father, a Kung Fu lover, who forced him to learn the art. Eric laughed, shaking his head, "This was entirely my own interest. I wanted to learn Wing Chun because it's popular. Once I told dad about this, he tried hard to find me a good master. Dad would not force us to do anything. However, once we decided to do something, he expected us to persist in our interest."

Nicole talked about her father's good sense of humour. "Although daddy seldom told us his hardships in the past, occasionally he would reveal what his life was like in the countryside," she said. "I remember once I asked him to beat a cockroach as I was really scared. He said, 'You're scared of cockroach? I used to eat it!'"

"I remember," Lesley added, "when I was going to take the college entrance examination, to alleviate my pressure, dad said our family still had land in the village and I could be a peasant if not a college student. I laughed, and soon the pressure was gone."

"Actually, dad is very humourous," Eric agreed. "He likes to tell jokes. However, if you don't know the context, you're not able to get the joke." Sometimes when Simon told a joke, his son was the only who got it and had to hold in his laughter.

Lesley remembered her father as both very serious and good humoured: "He was rational and careful, but he was also funny and playful. He would play some fun but risky games with us, such as bang snaps, which are similar to cherry bombs, and wax burning, a traditional game in Guangdong played often during the Mid-Autumn Festival when the kids melt candles typically onto a metal surface of a mooncake box. We had a good time playing these games together and we're all very close to dad."

"Every Mid-Autumn Festival, dad lit a dozen candles on the lid of an empty

❖ *Simon playing fireworks with his kids*

mooncake box, using his foot to hold the tin box in place. Then he flipped the lid onto his head. We all felt it's dangerous to play this inside the house, but every year dad wanted to perform his feat for us."

Nicole said, "Dad encouraged us to play and when he played with us, he was even happier than we were, especially when he played wax burning on Mid-Autumn Festival. Others warned us it's unsafe to play, but he didn't care and said, 'it would be fine. Let's play.' I still remember although the trick appeared dangerous, he would perform the feat each time flawlessly. I'm also the same way. Once I want to do something, I just do it, like enjoying the adventure of travelling to different places."

"Dad liked exercising," Nicole added. "Every now and then he would stand upside down at home." Simon's handstand left a deep impression on Lesley and Nicole. "It's so cool." Dad's devotion to exercise had strong impact on Nicole who exercises regularly and has a passion for sports.

When Simon was a child, he led his little friends to race on a field under the moonlight; when he became a father, he took his children to play fireworks. Behind his playfulness is a soft and warm heart. He is also considerate and started the family tradition of sending a handwritten card.

"I went to the US at 13. On my first year there, dad wrote a birthday card for me. Later I would receive his card every year," Lesley said. She had studied abroad from an early age. "These cards conveyed dad's values and ideals. Even though we're thousands of miles away, I would still receive dad's encouragement."

Simon sent a handwritten birthday card every year to each of his children. He knew each of them very well. The cards conveyed the same care and love while carrying a different set of wishes and expectations. Every word expressed Simon's philosophy and his wise guidance on life. But perhaps, most importantly, between each line was a father's deepest love to his children.

Simon's children inherited the tradition from their father and would also send a handwritten card to express their best wishes.

"My friends were very surprised when they first received a handwritten card from me. They never expected that someone still wrote out cards. However, it's a tradition in my family," Eric laughed, "I will send a handwritten card to my parents on their birthdays, and also on Father's and Mother's Day. My sisters and I would also send to each other birthday cards. I have kept all the cards. They are immeasurable treasure to me." On Simon's last birthday, Eric sent him a card with a superman on the cover. In his heart, his father was indeed a superman.

❖ *Simon drew a birthday card for his daughter.*

According to the Dharma Master Cheng Yen, parents' love for their kids is like the natural and non-stop flow from a spring. Simon's love for his children is tender. As a father, he is responsible and affectionate. Above of all, he trusts his children and communicates with them like a friend.

"Dad is reserved and doesn't express himself directly. He prefers writing his feelings down." Eric often received words from his father, which contained Simon's wishes for him, to be "staunch and pragmatic, brave and upright". Simon would put down his thoughts in a few words and shared them with his children. These words expressed the very basic principles of his philosophy.

"This year's Valentine's Day, I sent dad a card, saying he was my love in my previous incarnation. He replied with a long message, saying that 'you will meet different men in your life and each of them will teach you something. The last one that stays by your side understands you the most.' What he said went straight into my heart." Lesley said that her father had many traditional family values, but he was also open-minded and accepting of his children's choices.

Whenever Simon saw a painting that moved him, he would immediately share the impression with his eldest daughter who managed an art museum. He would take the time to select an appropriate calligraphy to match the design of her daughter's office and hang the painting with her. "You sit down to see whether it's about the right height," said Simon, standing behind the chair while looking like a big tree.

Simon enjoyed buying clothes and accessories for his children who all admired his taste. Lesley said that she still wore the cashmere her father bought for her ten years ago.

"He knew which style suited me. The colour and the style looked very good on me. It wasn't always a big brand. The clothes dad bought for me were of high quality and could be worn for many years."

"On my birthday, dad sent me a pair of earrings. It's really very nice." She added, "Look, this dark green jacket on me was also from dad."

Simon often said, "Harmony at home brings prosperity." As she's grown up, Chloe has deeper understanding of it. The close bond in the family is definitely nutured by the parents. "Perhaps some people would think my parents doted more on Eric as he was the only son in the family, but I never felt this way," said Chloe. Nicole also agreed. "I had some friends who felt neglected in their family because they're the daughters. However, in my family, dad treated us equally and was never partial to any of his children."

Eric agreed, "Dad was not the type of a traditional Chinese father who loved his son better than his daughter. He treated us all the same. Lesley and I were about the same age and often played together when we're kids. When we quarreled with each other, dad would not stand by my side even though I was younger. He didn't care whether I was right or wrong and said as a man, I should be more broad-minded. He was even stricter with me because I was a man."

"Dad has this one traditional family value: a man should work and support the family. Although mom also has many public services, dad still holds onto his traditional value. As the head of the family, he should bear the responsibility to take good care of every family member."

Mary supported her husband at every step. She started the business with Simon; later she took a back seat in business and became devoted to charity. She accomplished a great deal in her own right. Simon joked, "My wife's schedule is even tighter than mine."

Mary was awarded the 2018 Outstanding Businesswomen Award in Hong Kong. Simon went with her to the award ceremony. He said in an interview that he was even happier than his wife about the award. After 35 years' marriage, the couple still love and respect each other.

According to Confucianism, a family will prosper if the father is affectionate and the son is dutiful; the brother and the sister love each other, and the husband and the wife respect each other. If one can have a harmonious family, they will be able to cope with all kinds of problems in life. This is what a family should aim for.

Obviously, Simon has set a good example in this regard.

❖ *At a staff award ceremony, Simon wearing a traditional Chinese long gown, espouses his philosophy for succeeding in business: a philosophy resonating with traditional Chinese culture.*

CHAPTER

# VIII

## CREATIVITY

/

## DOING BUSINESS

# Sit by a Raging River, My Heart Is at Peace As the Cloud above My Head, at Ease

Simon has frequently remarked, "Business means busy thoughts". The statement may seem a mere truism; nevertheless, the four words capture the very essence of Simon's philosophy of doing business: making changes and adapting to changes.

The expression, "Thoughts" is a characterization of the creative, innovative mind, while the adjective, "busy", emphasizes a mind should remain active so that the innovations can be placed into operation. In Simon's words, doing business requires "emphasizing the reality and recognizing the changes": making modifications in response to changes in the environment. Guided by the philosophy, Simon's ways of thinking and decision-making often come as unexpected to others yet enables the keen business leader to succeed where others have buckled, succumbing to unforeseen change in climate and current.

Ronny Ho, a colleague who has worked with Simon for years, understands the daunting task of grasping the business leader's method: "I'm used to making judgments based on causal chain, one factor contributing to the next one thing, like the shape of a tree — bottom up, from trunk to leaves," Ho says, "However, Simon is very different. He relies on his instinct, as if he were jumping randomly from one star to another in a star-studded sky. The amazing thing is, his instinct comes from an unusually astute observation of the people, the market and the whole business.

Ho has worked in many large companies before he joined Suen's company. According to him, the business tycoons he has worked with all have different strengths in business management and operation; however, none was gifted with Simon's ability to survey the spectrum of opportunities that Ronny calls "a sixth sense" as Simon does. "Perhaps it's because he likes reading philosophy books," Ronny laughed. "Just look at how he wins a bet on a horse race. It seems random, and he cannot even explain why he often wins."

Horse racing is a Hong Kong obsession. In 1998, Simon officially became a member of the Hong Kong Jockey Club. His first bet on horse hit a big prize, winning him a fortune. "At that time, my head was all mixed up because I desperately needed a major sum of money to settle a payment," he recalled. "A friend of mine invited me to a horserace, feeling uncomfortable to refuse, so I went with him. I randomly wrote some horse numbers on the lottery tickets. To my surprise, every ticket drew a prize. The money I won helped me with the cash flow."

In speaking of his first experience at the racecourse, Simon used "When one door closes, another opens" to phrase the moment. He had won many big prizes in horse betting later; however, Mr. Wu Yan Wah, a frequent companion to the track, remarks that Simon knows nothing about horse racing at all, such as the classification, score, and all kinds of indexes and figures. Neither does he rely on the same kind of preparation that others employ in making a bet. "If Simon finds a horse and the rider in good spirits, he will bet on the horse. The only index he refers to is the odds. He never bets with the low odds even if he feels good about the horse. The return is simply not worth the risk. Thus, Simon often places his bets on horses that are considered by others to be unpromising, but ultimately, he emerges as the winner." Simon's unique betting style has never failed him in the thirty years of their acquaintance.

"Such a unique sensitivity, when applied to business, is undoubtedly a primary factor contributing to Simon's success — he always grasps the opportunity, seizes the right moment, and gets it right," said Ronny. "The change of his business from woven to print label exemplifies his vision. Before the change, over 99.9% of his business was in woven label: a business allowing him significant profits. If he was just an average businessman, he might have stuck to woven business without bothering to make a change. But he did not. He sensed a wind of change in the markets. To increase competitiveness, his company would need to provide one-stop services of woven and print label at the same time. With the strategy in mind, he began to re-deploy his business through strategic acquisition, talent recruiting and resource obtainment. When the new wave hit the market, his business went through the transition smoothly, thanks to his extensive preparation. Without his astute vision and timely adjustments, his business would have hardly sustained."

Wu and Simon, both horseracing fans, also worked in the same trade of label business in the early years. In the 1990s, they made a business trip together to Hubei — a province in central China — to recruit workers.

Even now in his seventies, Wu's memory still serves him well. He recalls vividly many details about the trip. "We took a green train — the slow, cheap and old-fashioned one — leaving in the morning and arriving at night. The weather in Hubei was so different from that of southern China. The weather was fine when we stepped onto the train, but by the time we got off, it was snowing heavily. Taking only light clothing with us, we had to bundle up in rented overcoats almost as thick as a quilt with cotton wadding. After that journey, we still needed a bus ride to our destination as it was a remote city. At that time, cities in central China were less developed. As a result, public transportation was dilapidated. A bus with a damaged frame would still be permitted to travel the road. Even so, we took the bus, striving to earn a living, we could not afford to be fastidious. Besides, we were able to adapt to changes very well. Simon had a habit of reading at that time already. Whenever he wasn't busy, he was reading a book. I didn't pay attention to what he's reading exactly, but I think those books must have inspired him."

"It was in 1995, if my memory didn't fail me, when he told me he wanted his company to become No.1 in Asia and even go global. At that moment, I thought he was talking nonsense, although I dared not say so to his face," Wu laughed. "label factories were all over the streets then, so there was very intense competition, and not much profit could be earned. It was not easy for a factory to survive, not to mention going global." Then, in the same breath Wu said, "But Simon possessed resolution and audacity. Having a long-term vision, he took actions quickly. When he

❖ *Photo taken in Harbour Plaza in Yau Ma Tei in 1995. For the first time, Simon mentions to Wu his plan of going global.*

set a goal, he would go for it. You might not know this industry well enough to judge his success, but with my experience, I think it is fair to say he has created a miracle in the apparel label industry that he can achieve such success."

"He has the guts to innovate by taking a road less travelled. Not only is he bold, but he is also responsive to change. There is one thing I admire about him: whatever he wants to do, he will do it and make it. Back then, his label factory had only a production capacity of several millions, but he dared to take an order of ten times bigger, and he was able to handle it," said Wu. Believing himself a more prudent and conservative person in business, Wu used a vivid metaphor to describe Simon: "When there are ten bowls of rice on the table, I would eat just one bowl, but he would take all of the ten and finish them. That's the way he does business. I really admire him."

Not coincidentally, the founders of Amazon, Jeff Bezos and Blackstone, Stephen Schwarzman, are also "high achievers motivated by an appetite" for success, continually driven to pursue more and higher goals when starting their business. Such "good appetite" was believed to be the momentum for the unceasing growth of the companies.

"He is a visionary guy. When there's a conflict between long-term strategies and short-term interest, he knows how to adapt and make choices. Once a decision has been made, he will act resolutely. That's his vision, and his guts." Ronny commented. "Deciding to end partnership with UPDL is one such example."

In the 80s and 90s, the suppliers hadn't set up direct partnership with brand owners and had to rely on agencies to get orders. As a leading agency, UPDL had been a long-time partner of SML. However, the late 1990s saw the change in the operation mode of the industry. Brand owners and label suppliers started to avoid agencies, contacting one another directly.

"We had been successful in gaining orders from GAP. Seeing that SML did not need an intermediary to form an effective relationship with GAP, those agencies started to treat us as competitors instead of partners. UPDL was one such agency." Ronny said, "The management team discussed how to address our relationship with the potential adversary. Personally, I preferred a more conciliatory approach. However, Simon sensed the future direction of the industry and took the initiative to call a halt to the cooperation. He saw the necessity for our company to take the lead. Of course, the rupture in this relationship had immediate repercussions to our business, but the decision served as an important step contributing to our long turn growth. It enabled us to seize on some golden opportunities to boost our business, and overall, deepened our bonds with customers. Simon has the unique ability to both intuit the changes within the industry while remaining sober minded throughout the decision-making process."

Maybe his "correct sense" in other people's eyes is exactly his sober mindedness.

Simon usually says that a whole picture in mind is a prerequisite for making a decision. He believes that confidence comes from accumulated experience and a full understanding of the game, which involves psychological analysis, profit evaluation, and risk management. He also needs to work out solutions to the worst possible situation. For example, Simon once decisively terminated a negotiation with a well-known Japanese corporation. "As a small shareholder, the company asked for a big shareholder's right," Simon said. "Our benefit would be significantly dwarfed by the risk we bore. If we did introduce the Japanese side to our board of directors, it would be counterproductive. Our hands would be tied, and the gains couldn't make up for the loss. I prefer to establish a horizontal cooperation which might work better." Given the complexities of the decision-making procedure, the Japanese delegates were not able to make any adjustments to the goal, but personally, they were deeply impressed by Simon's vision.

Simon's vision embodies the wisdom of *The Art of War*, a Chinese classic. In this book, it says that the fate of a small stone in a river depends on the flowing water, so the stone should make full use of the strength lent by the current flow and simply follow it. A swift bird can easily get its prey because it can seize the right moment. The victors in a war can grasp the "trend" and the "moment" at the same time. This is where Simon's sensitivity lies. He gets his hands on a trend in world politics, economy, businesses, and industries. For another, he can always hit the right moment. That's why invariably he wins out.

Although poles apart, running a business and running a horse have much in common. For horse racing, one's intuition may help in measuring the condition of a horse and its rider, but Simon also holds high the "small-chance-for-high-profit" principle, placing his bets where he may achieve a maximom profit. The decision-making process exemplifies a combination of sense and sensitivity. Similarly, as a businessman, he boasts a result-driven mentality, focusing on the return on capital, assets, and effort.

"When we are making mid-term and long-term plans, we typically take into account past performance and future prospect: then figure out some feasible plans, ranging from the conservative to an aggressive approach. But our boss would refuse the rigid framework, viewing the situation from a higher plain. At the first glance, you may think that his proposal wouldn't work, but in fact, it is exactly the core that he grasps, and in his proposal, everything else just goes around the core and serves it", said Ronny Ho, who knows clearly about Simon's "outcome-driven yet outcome-transcendental" mentality.

Linus Cheung, an old friend of Simon's, has remarked that Simon was able to transcend the restraints of a specific business deal, achieving a panoramic view of the situation. This wisdom is not made for everyone: he confesses.

"Being astute and capable are indispensable qualities for a businessman, but these qualities alone cannot guarantee the survival of a business. Capability is not tantamount to wisdom. Simon is a man of wisdom, I have to say. And his wisdom is even rarer in the world of business," said Linus Cheung.

Linus is nicknamed the "king of worker" and has worked in a range of fields including as Deputy Managing Director of the Cathay Pacific Airways, CEO of the Hong Kong Telecommunication. He is also an entrepreneur and has become acquainted with Simon due to a shared interest in art.

Linus speaks highly of Simon's "advance-and-retreat" philosophy in business.

"For advance and retreat, it is easier said than done. In running a business, you must have your own judgment on whether it is the right moment to take action, and whether it is the right viewpoint to analyze the business."

Linus has also in the past been a radio host and speaks eloquently of his friend's strategy: "Simon's success relies on his vision, judgment, sensitivity, and, of course, on his advance and retreat tactic. He knows clearly when to adjust his business, when to use technology to help his business, and what direction his business is moving for. You may already know that he sold the SML Tower. That's typical of him!"

In 2020, the global impact of the Covid-19 pandemic challenged world economy and brought changes to individuals and businesses alike. For example, platform economics, new retail, and WFH (work from home) have all become the common work-related practices. As a result, the demand for business offices has dropped by at least thirty percent due to the popularity of WFH. By contrast, the rapid rise of new consumption patterns has brought opportunities to the businesses that provide technological support to the new consumption pattern.

"Riding on this wave means that you catch a new opportunity for development."

Simon caught the moment and decisively sold the SML Tower. However, his act was met with suspicion due to the gloomy economic environment at that time. But Simon maintained his composure, saying that "we should place eyes on the trend and the long-term development, and wisely allocate resources so that we will not miss any golden opportunities in the future."

Upon reading a news report about Simon's selling of the SML Tower, Linus sent a message to him — "Truly strike the balance between advance and retreat. Salute!"

In the same year, Simon transferred the leadership of the corporation to the team he built. "I stepped back, not for enjoying the retirement life, but let the new generation to give full play to their talents."

"The company is like his child, whom he carefully raised up. No one is more emotionally attached to the company than he is," said Ronny, "But he is willing to loosen his grip on it, allow his team to run it, and introduce external capital to it. All he did is for the benefit of the company." Ronny is fully aware that Simon's retreat is for the company's advance.

Venerable Master Hsing Yun said that life is like tango, and one should go forward and backward. That is the essential wisdom of advance and retreat. To learn to advance is not easy already, and to learn to retreat is more difficult. Linus regarded Simon's understanding of advance and retreat as outstanding, as he said: "Simon is very generous and forthright with his friends, and this may have something to do with his great passion for the book, *Water Margin*. However, it does not mean that anyone who reads this book can learn something from it. It depends on whether that person has exceptional understanding of its underlying principles."

Simon may speak in a manner that sounds impractical to his listeners, but there are reasons behind the seemingly abstract tone. In the Wei and Jin dynasties of China (AD 220-420), scholar-literati tended to refrain from talking about things concretely. Instead, they discussed a lot about the very core and the nature of phenomena. When such a way of speaking is applied to business running and horse betting, it sounds as if one is being advised to avoid a concrete analysis of data and information. Yet in fact, the point is to guide the listeners to look at the forest, not merely the trees. Or, in other words, one should assume a proper attitude, adhere to a correct philosophy, and hold a vision apart from concrete circumstance.

This is precisely the essence of Simon's remarkable success. He has truly grasped the nature of betting, either for horse or for business. Seen from results, low chance should bring in big return. Seen from attitude, one should be bold enough to lose and win, and know when to advance and retreat. A line from the movie, *The*

*Godfather,* reads that those who can use half a second to penetrate into the nature of things must lead a completely different life from those who still cannot see clearly the essence of matter even with a lifetime.

Simon, of course, belongs to the former. An average person may see his words and deeds idiosyncratic, but impossibility for others often becomes Simon's possibility. A sentence from the Chinese classic, *Tao Te Ching,* can best describe Simon — "The impossible beyond the impossible lays the gate leading to marvels".

Simon once learned Chinese medicine, studying the Ba-gua for a long time, and is also a huge fan of traditional Chinese art. In fact, underlying his business philosophy is the Chinese traditional wisdom.

"Doing business is like composing a painting. Both illustrate the fluid interplay between the static and the dynamic as well as the abstract and the concrete." This may sound unimaginable but not when Simon drew on a concrete example: "It's like the interdependence and interaction between traditional business and technology products. Traditional label is the vehicle for technology innovation which in turn becomes the catalyst for the transformation of traditional business. Traditional label business alone does not work without technology, nor does it the other way around. Our business can't sustain without a balance between the two. They must be well integrated to ensure long-term business development."

"Although it may be highly profitable, the business unit should be cancelled or discontinued before it becomes obsolete. We should be forward-looking by embracing the latest trends in the digital manufacturing era, such as Internet of Things (IoT), Artificial Intelligence (AI) and 5G Process Monitoring in Manufacturing Systems."

Simon has a typical response to his colleagues' skepticism about a proposed change:

"There's nothing wrong to follow the logical and scientific model of Western enterprise management such as SWOT. However, a form of analysis, based solely on rationalism and logic, often leads to conflict. These divisions make it hard for a team to achieve consensus. This is when traditional Chinese wisdom comes in handy. Traditional Chinese culture teaches us the dialectical thinking of yin and yang. It is similar to Kung Fu: you need to channel all your force when throwing a punch and be decisive about when to attack and when to retreat."

"For example, when we started our business, the looms were very expensive. However, we had to buy the equipment because it was necessary to increase productivity as there was a high market demand. It didn't matter that the investment cost was high, as long as it enabled us to grasp the rising market opportunity. As the market focus changes, we have to respond quickly even if it means a short-term loss for our business. Numbers alone cannot explain the change in our strategy. The market should be the key determinant. Responding to the market is essential for business growth. Our production should be determined by the market, not the other way around."

Simon's philosophy for doing business does not come from adhering to a fixed set of rules but through his thorough understanding and flexible application of core principles. "Although numbers are fixed, the business — just like the painting — is not static; therefore, a static way of thinking will not work."

Simon has developed a regional label manufacturer into a global supplier of retail technology solution with a presence in over 30 countries. This successful transition is not possible without his notion of "scroll painting" whose composition is sensitive to different times. Nevertheless, one thread remains unchanged: his original aspiration to make his business a great success.

From the outset Simon has been clear that managing people remains the key behind a successful business.

"It's all about managing the people and the jobs they do. They are the two sides of a coin. Effective management of people ensures a successful completion of jobs. Otherwise, the jobs can never be accomplished." After a pause, Simon added, "It's true that we need to make plans, budgets and KPI. However, according to my own experience, I would say it's equally true that only when we manage the people well can our company enjoy long-term development."

"Business management is all about the three 'qi' — 'Renqi', 'Zhiqi' and 'Shiqi' — they are all related to people. 'Renqi', or solidarity, means that all the employees are of one mind and share common goals; 'Zhiqi', or aspiration, refers to their determination to achieve the goals; 'Shiqi' stands for morale — the company should boost their morale by making clear the reward once a goal is achieved."

Simon has his own unique ways to boost the three 'qi'.

"We have a long tradition of holding different activities for our employees in Dong Hing from the sports day and Mid-Autumn Festival gala to lectures and seminars for the staff." Ronny, who has worked for Simon for many years, says, "On a regional and international level, the activities can instill in our employees a sense of belonging, establish a shared identity, convey the vision, ideal and mission of our company and promote its development."

Simon often says, "We cannot unify people's minds, but we can unite them with shared goals."

The key lies in channelling people's spirit and energy towards achieving a unified vision. Simon's idea is rather similar to that of Ren Zhengfei, the founder of Huawei. Both believe that the managerial personnel should often dine together. Simon has not been particular about food. Only in recent years did he begin to develop strong preferences. However, he possesses an acute understanding of how the Chinese dinning culture functions. To some extent, the art of management is analogous to the art of dining. Simon has been practicing this idea of "uniting people through dining together" in Dong Hing, and he has not deviated from this core principle even as his business has grown to being a global entity with a presence in more than 30 countries worldwide.

At the annual gala of the Group, the management from all over the world gather together. Some regional teams put on talent shows; other staffs share their stories. Simon gives a speech at the gala every year, describing the "painting" in his mind and sharing the long-term, mid-term and short-term goals of the Group. The event represents more than an annual dinner. By bringing employees from different places closer, the gathering unites their hearts with shared goals and missions.

Every year, during his overseas business trips, Simon invites local management team to dine with him. Even though he visits many of the places once every few years, he knows very well the strengths and weaknesses of the local team. Such insights come not only from the work reports at meetings but also from his observation at the dining table. He outlines his view of the Group direction and local positioning to inspire them, broadening horizon and having a fresh thinking.

Simon is comfortable at the head of the table, knowing how to galvanize his colleagues and to enliven the atmosphere. Sometimes when meal reaches a climax, he would recite a poem. In 2019 when dining with the managerial team in Spain, he composed a poem impromptu:

*A gaze into the Mediterranean Sea,*
*Off which comes a gentle breeze,*
*Awakening green willows and red leaves,*
*United as one we shall be.*
*My heart is lifted up by mountains great,*

❖ *The opening ceremony of the first Dong Hing Sports Day in 1995: Simon following the staff team to enter the venue*

❖ *Simon at the Dong Hing Mid-Autumn Festival gala, warmly welcomed by the colleagues*

*❖ Simon (front row, second from left) dining with the management team of the Hong Kong company in 1997*

*❖ Simon (third from left) and his Asia team dining with the Europe team in Portugal in 2019*

*Against numerous hardships and pain*
*Forging ahead as we are brave,*
*Together, a bright future we will create.*

"We should let our employees feel they have a promising prospect. As long as the goals are achieved, all their efforts will pay off." Over all these years, through his own unique ways, Simon has boosted the morale of his teams, establishing an atmosphere of trust. "This is where the motivation and positive energy of an enterprise culture lies."

Many members of the core management team have met Simon for the first time at the dining parties. The current Chief Administrative Officer of the Group, Cherry Cheung was one such individual.

Cherry joined the Company in 1990, starting as an intern. Three years later she met her boss Simon for the first time at a dinner banquet. Cherry served in various capacities at the many different departments of the Group and, due to her diligence and her down-to-earth character, had, over her years of service, become entrusted with greater responsibilities.

Once, Simon visited Cherry at home and saw she lived with the other six family members in a cramped 300-square-foot public housing flat. He appreciated her sense of duty as an eldest daughter, fulfilling her obligation to help shoulder the family responsibilities. He decided then Cherry must know how to pinch and scrape and that the background was the proper soil for yielding a dutiful and responsible employee. The Chinese believes that each place nurtures its own inhabitants. Simon has an incredibly discerning eye in his judgement of his employee.

"Once we were on a business trip overseas, and in order to save money, she arranged for me to take a bus. But it took us quite a long time to get to the bus station, not to mention that we had to drag the luggage throughout the journey." Simon smiles every time he talks about the experience. "She spent every penny of the Company's money very carefully. I know I can count on her." Needless to say, he appreciates her deeply.

Cherry wears short hair and has a comely face. "It was a trip to Düsseldorf. At that time, we didn't yet have an office in Germany, so I arranged the itinerary myself. I didn't rent a car as the airport bus stop seemed very close to the hotel, but we walked a long distance. Unfamiliar with the way, dragging our suitcases while looking for the hotel and worrying about missing the lodging, I was getting more and more anxious. Fortunately, Boss wasn't mad at me."

Simon has paid home visits to many employees. "This is one of the things that sets him apart from other bosses," says Cherry, "He often says that his thousands of employees around the world represent thousands of families, and that gives him huge responsibilities to run his business well. On the one hand, you can see his big vision and great wisdom in running the business; on the other hand, you can also see his attention to small details and needs, such as the living environment of his employees. The canteen in Dong Hing is an example. It was renovated just the year before last at Simon's request. You rarely come across a businessman like him who displays such an interest in the welfare of his employees."

There is a saying about the wisdom of doing business, to the effect that people who are meticulous usually have a poor vision, whereas those who have a good vision are usually inattentive to details. You are good only when you can do both well, which is exactly why Simon is skilled.

"Our staff can enjoy their hometown cuisine in the canteen and feel relieved of their homesickness. It is only when they can work with a peaceful mind that the success of a business is assured." Simon recognizes the importance of the business to expand globally while showing an equal concern for the livelihood of every employee.

"Simon committed a great deal of time and effort to building his team in Dong Hing. He could be quite strict at work, but he also got along with his colleagues after work. Often, he took them to sing Karaoke and dance together. He always bought a cake for a colleague's birthday. He cared about others from the bottom of his heart. As the General Manager of the factory in Dong Hing for six years, Cherry has seen Simon's care for his employees.

As Simon sees matters, a business must be profitable and achieve a steady return, but for that success to be sustained, a company must be built on a "people-oriented" culture. "As a boss, I need to be concerned about both the tangible and the intangible, achieving macro and micro-levels, as well as examine situation from bottom up and top-down perspectives. Only then are we able to clearly see the circumstances and conditions facing our company."

"In the early days when we went to attend overseas exhibitions together, no matter how busy the itinerary was, he would squeeze some time to take colleagues around for some sightseeing and shopping. He knew very well how to be in the shoes of other people. Of course, since our business is closely connected with retailing, this can also be seen as learning about the market. He would also pick some gifts for everyone," Cherry recalls, "In 1997, he led a team to Italy to attend an equipment fair. That was my first business trip with him. After we finished our work, he took us to a square, a landmark, I think, where there were a lot of white doves. We took photos, then went to a famous commercial street. I remember he bought me a notebook. It wasn't anything expensive, but his thoughtfulness was really appreciated."

Li Hanwei is currently in charge of technical engineering for looms of the Company. He is not tall and has small eyes but a very smart look. He joined the Dong Ying joint venture in 1983 then Dong Hing where he has remained ever since. In the early days of the Group's expansion of overseas production sites, he provided technical support and took charge of the installation of looms. As a part of his obligations, he often travelled throughout the world.

"Once the Boss came back to Dong Hing and asked me to come to his office. I thought he wanted to assign me some work, but as soon as I walked in, he handed me a squarely wrapped box. When I opened it, it was a 'Quick Translator'. He knew my English was poor, and said it would come in handy and would be a useful learning tool. He was really thoughtful."

When he was running the Dong Hing Factory, Simon wrote cards to congratulate colleagues on their birthdays. With the popularity of smartphones these days, he sends more text messages. "Earlier on my birthday, Boss wrote a poem to congratulate me," Ronny said. In his spare time, Simon also posts messages to his colleagues, sharing his views on certain issues. "I often receive messages from Boss about his thoughts and views on the Company. I compare his thoughts with mine, discovering new ways to look at things. These exchanges are greatly inspiring," says K.C. Lau.

When leading a team, Simon often talks about how to manage the members: "We should teach and guide them, rather than getting angry and reprimanding them whenever a problem arises. If you haven't taught them, how can they be able to learn without making any mistake?"

Simon was not only the Chairman of the Board but also the Principal — he ran a school in the Factory, where he gave colleagues professional training. He taught his managers the "doctrine of the mean" to inspire them to think in management terms. His teaching was to the point: "The doctrine of the mean is not about playing safe but about being impartial: it is not about doing nothing when you see a problem, but about being audacious in reforming, exploring and expressing your own views."

The management team of Dong Hing, who Simon calls his "business backbones," have become entrusted with positions of leadership in the Asia-Pacific region, thanks to Simon's early efforts to prepare for the rainy days. His emphasis on "teaching" may have its roots in his early experiences as a teacher when Simon imparted his drive and determination onto his students.

The important role of a teacher is to share knowledge and experience, so that students can continue to improve and grow in their learning. It is the wish of a teacher to guide his students to be better than himself, while adopting the mindset to help others succeed that allows for Simon to be both an entrepreneur and a teacher willing to help and support his subordinates and team to achieve self-improvement. In the words of Simon, "You'd better be better than I am."

Teachers must be skilled communicators: they should know how to teach and share their experience to inspire others.

K.C. Lau, who often accompanies Simon on overseas business trips, takes out a photo of Simon writing in his notebook while Lau looked on intently.

"This photo was taken during a meeting in Spain. At that time, I was discussing with the Boss whether to start a business in Portugal and if we did, how we should achieve the task. He took my notebook and wrote down these words: 'To make oneself stronger by striving constantly and by overcoming oneself'. I wasn't CEO back then," says Lau while taking off his glasses and depositing them on his desk. On the top of his laptop is pasted a label of a flying fish.

K.C. continued: "To me, these words have two important meanings. First, we must make constant efforts to learn, improve and strengthen ourselves throughout our lives. Second, we must overcome ourselves before we can overcome challenges. They are not just about how to do things but also about how to become a better person."

Simon's way of teaching is not simply to guide the management team on a step-by-step solution to a problem but more importantly, enlightening and inspiring them.

"One of the keys to good human resources management is to build talent force," Simon said. "This includes the training, promotion, recruitment and motivation of talents." When he spoke, it appeared as if Simon was once again the young founder, seeking out talents so eagerly.

Oftentimes, he shares his team-building experience with the management team. Once, Simon wrote a message in person to the new head of a production site, which reads: "the success of an enterprise counts on the qualities and capabilities of its leader. *The Art of War* also says the commander must stand for the five virtues: wisdom, sincerity, benevolence, courage, and strictness. Those are the qualities of a talent. I hope you can be patient, tolerant, and protective when training your staff so that more talents will be recruited and cultivated. A leader should look up, so to speak, at his subordinates, listening humbly to their voices — and most especially to their complaints. The principles of a benevolent leader first spoken by Sun Tzu comprise truly the philosophy of people-oriented management."

Simon observes: "When talents are recruited, and the mechanism for talent management is also in place, your team will be strong. Work is easier done that way. People are important. When the right persons are placed in the right positions, they can make full use of their talents and give full play to their wisdom as members of our team."

Simon knows not only how to train his staff but also how to assign a job to the right person. Wu Yan Wah, Simon's long-time friend, said, "A weak commander relies on strength. A mediocre commander uses his wisdom, while a strong commander manages his people."

Simon knows how to teach and to manage. "Even a careless employee can be trained by him to handle some meticulous work very well," Vera Chan, his secretary for over three decades, said that Simon uses personnel in a unique way. Simon assigns a job to you, not because it falls within your duties, but because you are the right person to carry the job out. If you are good at the task, then you should be the one to perform it. Conventional ideas or rules do not restrict Simon. He believes that everyone should be able to give full play to his strength in the right position. "Therefore, you might be recruited as an accountant when entering the company but end up being a member of the marketing team."

He also understands the duties of some specific positions uniquely.

Simon said, "Business growth largely depends on the market; therefore, a regional head should have sales experience and understanding the market, and the ability of respond to the market trends. However, heads should not lack finance knowledge and have weak control of the cost. Therefore, a combination of marketing and finance is ideal. Surely, as technology innovation has had a substantial impact on our business, it is also vital for management to understand technology."

Simon often reads from the writings of Mao Zedong, finding a favourite quote: "senior cadres are of paramount importance in the implementation of policies." "The new leaves drive away the old ones; the waves ahead give way to the ones behind." For Simon then, "It is only natural to make steady progress. The company's moving and changing, so does the management team."

Every human being is very unique. It takes the highest wisdom to form a team and encourage the team members to reach a consensus. That is another trait of Simon: to achieve agreement while leveraging the individual skills of each employee.

When the company grows larger and its business expands to more regions, opinions might differ at times between functions within the region, among various regions, and even between regions and headquarters. When a consensus cannot be reached, the problem would be reported to Simon, who personally handles those issues.

"We'd better figure out the fundamental reason for the problem first. Every one of us wants to do things right, so we have to be more inclusive. But we all occasionally have blind spots, and these may prevent us from grasping the whole picture, so it's important to have a broad vision."

He pointed to the files on the desk and said, "In some regional teams, they follow the rules and regulations to handle matters concerning employees who have been working with us for a long time. That is, of course, correct. But we have to communicate first, and then try to solve matters. Of course, harmony is important to business success, so those matters should be handled more flexibly to avoid disputes. How you handle a matter is as important as how you make a decision. At times, a hard or soft stance can be taken, but you need to have your own style in terms of methods and communication techniques, and with confidence, to solve and resolve the problem. This is the art of management."

In Simon's bedroom hang ten long scrolls written by Professor Jao Tsung-I on Confucianism, Taoism, and Buddhism. He stares at those scrolls every night, pondering their meanings carefully.

"There isn't one formula appropriate for managing all enterprises," said Simon, "To manage an enterprise, we need to learn constantly and live with trial and error. In the end, we will experience what Taoists describe as 'governing by doing nothing that goes against nature.' The principle translates in business as when you have reached a certain level in management, you need to manage nothing instead of everything."

William Foote Whyte, a reputed sociologist in America, once observed the most challenging thing in the world is to hand over the work you are really good at to others, and then watch them messing things up while remaining calm and silent. Simon entrusts the business wholly to his team, no longer attending meetings on the daily operation. Knowing well that a decision could be made at the moment he enters the meeting room, however, Simon would rather let the team discuss and reach a consensus on its own. He just sits at the long desk in the hall and practices Chinese calligraphy, while applying a style of Chinese calligraphy, Xiao Kai, to transcribe gracefully and freely the *Tao Te Ching*.

"We all make mistakes. What matters is that we learn from our mistakes and grow from our failures," Simon said. "If we made no mistakes at all and excelled in all kinds of work, we would make a huge load of money. However, no such thing exists."

In Simon's opinion, a leader should be receptive to differences and is indeed wise.

"If his colleagues locate a better job, he never stops them from leaving. When he started the woven label business years ago, Simon hired laypeople, teaching them the fundamentals of his trade. Invariably, some would resign to start their own companies. Simon did not take the decision badly, understanding the need of the individual. After their resignations, those colleagues still came back to visit him. As a boss, he has achieved great success indeed besides business." Cherry Cheung, a long-time employee, remarked.

As a result of his years of experience in business, Simon has learned the advantage of putting himself in others' shoes. He is considerate of his colleagues and his business partners as well. His method of problem-solving typifies the win-win strategy. If a win-win situation is arrived at, the business will continue and grow. As a result, both parties will accrue benefits. The movie, *The Grandmaster*, shows how the mastery of Kung Fu is realized in three stages: being, knowing, and doing. In other words, a master must know himself, see the world, and ultimately passes on what he knows to others. As a martial artist, Simon practices these principles in managing his daily life and business. If he only knows himself well, he has no vision of others and the world. If he has seen the world and can pass on what he knows to others, that means he has a great vision. He can consider the world and all other people in it.

Cherry Cheung recalled, "Mr. Kuoni from Muller in Switzerland often visited Dong Hing. My boss, Simon, was a gracious host, even taking this business leader to the dorm to observe the staff's living environment. It might have seemed that he was being too honest, but this kind of honesty only deepened the trust between the two. In the 1970s, it was not easy for a manufacturer in Dongguan to establish favourable cooperation with well-known equipment producers in Switzerland. He succeeded in achieving that objective because he could think in the other party's shoes and have a vision in mind for their cooperation. His honesty brought him success."

"Whenever Simon visited Mr. Kuoni in Switzerland, Mr. Kuoni would entertain his honoured visitor at Simon's favourite restaurant, even booking his friend's preferred table. In fact, Simon has maintained an excellent relationship with many suppliers in the industry, such as Mr. Vaupel, Mr. Kuoni, and Mr. Saporiti Giancarlo from MEI in Italy. They have become friends after cooperating in the business."

While talking about Mr. Saporiti, Simon remarks that the two leaders share in common the value of friendships and recognize the importance of creating positive relationships. "I once visited him in Milan. He was so happy when I wrote 'You Peng Zi Yuan Fang Lai, You Yi Yong Heng' in Chinese ('How happy we are to have friends from afar. May our friendship last forever.')," Simon said.

MEI, headquartered in Milan, is a renowned manufacturer of woven label equipment. "Saporiti is the boss of MEI. He is not in sufficient health to afford long flights, so I go visit him occasionally. He has an amazing sense of fashion trends and can design and develop new products and equipment accordingly. He even invented many of the materials used for the products." Referring to his partner, Simon offers effusive praise.

Cherry used to manage procurement department, notes that in addition to those overseas suppliers, many local suppliers also respect and admire Simon. Cherry remarks: "After all, not everyone can achieve what Simon has accomplished. He is so capable that he has guided a Chinese company to such a position of leadership in the labelling industry that very few people can manage doing it the same way, especially when we know that this industry was previously monopolized by American companies."

Many of his business associates have become Simon's close friends. In addition to Raymond Leung, Susan, a close friend, who worked in Esprit, says straightforwardly that she admires his persistence, confidence and sense of responsibility. Simon's elder daughter was about to go abroad and study in America when the two companies started their cooperation. "Simon's strong responsibility to his family is extremely hard to find nowadays. He went to America with his

❖ *Mr. Kuoni from Muller in Switzerland visited Dong Hing looms workshop with Simon (left).*

❖ *Simon visiting Mr. Saporit Giancarlo, founder of MEI, in Milan, Italy in 2011. He wrote "有朋自遠方來，友誼永恆" in Chinese ('How happy we are to have friends from afar. May our friendship last forever.') Despite language differences, they shared similar wishes.*

daughter, looking for a decent school and even flew to America to visit his daughter every few weeks after she had settled down."

One can only be a good merchant if he has a complete personality. With his sincerity and honesty, Simon earned not only business success but also many friends across a wide array of fields.

Even Simon's competitors also admire him.

Matt Matsuo, an experienced expert in the industry of labels and accessories, served in YKK Osaka branch in the 1960s.

Matt was later assigned to America to develop business; he moved to America with his entire family in the 1970s and joined ASL in the 1980s. Asia was by that time emerging as an international garment sourcing centre, and it seemed natural for Matt, a leading specialist from Japan, to take charge of business in the region. As a part of his professional obligations, he flew periodically to Hong Kong to oversee the business development.

Matt said: "ASL opens Distribution Centre in Lai Chi Kok, close to the headquarters of Li & Fung Trading. There weren't qualified manufacturers for labels and tags in Asia those days, so we used to ship all products ordered by US retailers to ASL HK Distribution Centre, from where we distributed to countries throughout Asia."

On a bright afternoon in the early autumn of 1990, Matt was in the Kowloon Shangri-La Hotel, relaxing while enjoying his afternoon tea. It was at this very moment that he glanced at a magazine on a neighbouring table, recognizing immediately the handsome young man on the cover. It was Simon who had visited ASL five years ago. "His wisdom," Matt recalls, "and his profound understanding as well as incomparable passion for label industry left me with a very deep impression."

Matt immediately found Simon's number at the end of the exclusive interview. An hour later, Simon showed up at the gate of Shangri-La Hotel. The friends separated after all these years gave each other a big hug.

In the 1990s, ASL had been acquired several times, and Matt worked for different American label companies where he developed a healthy relationship with Simon. "I remember visiting Dong Hing a lot back then. I once gave some speeches to the employees, introducing the sourcing structure of American retailors and the constructive way to deal with these customers."

Matt joined R-Pac, one of the primary suppliers in the global label industry, in the year 2000.

❖ *Simon and his wife together with Susan and her husband. They had been friends for many years because of business cooperation.*

❖ *Photo taken in Matt Matsuo's home*

"As he has built up label business so fast and so big, he has also set up sales and marketing network with the same pace and scale. During my R-Pac period, SML was a good competitor."

When they first met in 1985, Matt found that Simon was an individual with a strong personality. However, the business leader also possessed an outstanding vision. "Under his visionary leadership over the past three decades, SML became the fastest growing label manufacturer in the industry with the most effective machineries to support highest yield of production," Matt says with admiration.

Having practiced Kendo for many years, Matt often practices with Simon, who is also keen on the Ba-gua Kung Fu. They set business aside on the court and focus purely on one another — both as an opponent and a friend. In the early days, Simon visited Matt every time he went to the States. Matt states outright that as someone who practices Kung Fu: "Dedicated, disciplinary, future-oriented, fighting spirit, all describe Simon's character. Thus, I believe that SML under his leadership could take over Avery's business and become No.1 Company globally."

Three words ending with "-ity", form crucial components in Simon's business motto. If running a business is about creativity, and handling clients is about sincerity, then employees should be treated with cordiality. Many employees view SML as a company with a human touch and willingly share their opinions on the company operations. Even after leaving the company, his former employees stay in a healthy and long-lasting relationship with Simon. He remains for them not only a passionate boss who dares to be creative and keep pace with times but also a wise, brave, and benevolent leader.

"Back in Dong Hing, it was completely normal that meetings were still ongoing at 11 PM or 12 AM. He didn't need to sleep when he was in the zone. We would all return to the dorms to get some sleep after long meetings and inspect the factory in shifts. But, whenever we were in the factory, he was always there." Sun Heyi, who worked as the General Manager in Dong Hing, recalls.

"We were on a business trip to Japan once," says Patrick Lau, who partnered with Simon at the early phase of his entrepreneurship. "It was in the middle of the night when he suddenly sat up and talked about his business plans and those great arrangements for the future." Patrick shares this anecdote to express Simon's commitment to shaping and achieving a company mission. While two later went on their separate ways, they remain friends who meet whenever there is the opportunity.

"I met the Boss at the Spring Dinner for the first time. He was probably the most energetic person I've ever seen — He was walking around all night and clinked glasses with each colleague present. It was almost as if he had never felt tired." K.C. Lau pauses momentarily, "He is an extremely good and fast learner. When we were in Israel learning new digital printing technology, he mastered the key features of the new techniques and figured out how the application could be transformed into businesses. It impressed me that someone could actually put so many thoughts together to form an initial plan so quickly." Lau has an IT background and regularly would go on business trips with Simon where he gained a deep admiration of Simon's capability as innovator and leader.

Simon's special drive "originates from mister's heart and persistence," explains Dr. Jeremy Liu, a member of the research and development team. Jeremy often greets Simon as "mister" as a sign of his warmth and affection for his boss and mentor.

Jeremy said: "I met mister for the first time after joining the company for two years, attending a Monday meeting with management. After hearing senior executives' reporting, mister summarized all key points mentioned and instantly planned the next-phase solutions accordingly. Sitting in my chair, I was stunned by how insightful mister is and how great he is at finding and solving problems."

He describes Simon as a leader who holds technology and innovation dearly. "What makes SML so successful is mister's commitment to technology innovation. Think big! Think long! This is what in mister's philosophy has inspired me the most."

Cherry Cheung, who has been working with Simon for many years, says except the ability to 'Think big! Think long!', "his wisdom is represented through his reverse thinking capabilities. It's his style to go for a unique track instead of taking the regular and common ones. Such a style obviously entails risk."

Invariably, Simon reminds his team to pay attention to potential threats. "An opponent may attack from all directions," Simon says, explaining the importance of remaining on guard. However, he continues to be bold, ready to seize on the next available opportunity. Accordingly, Simon is, in his own words, "a gambler kind of guy". For him, many things in this world are like gambling. The greater the reward is, the greater the risk is. No one can win for a lifetime. Simon reminds himself and others to be on the lookout for the weaknesses of human nature while preparing for the inevitability of loss and failure. Only by overcoming hardship, can one be "the biggest winner of life".

Cherry said, "When the result is worse than expected, Simon accepts the weight of responsibility. He always has his approach to handle adversity. Perhaps, because of his childhood experience and his training in Kung Fu, he is wise and brave; his experience gives him the courage to face any outcomes of his decisions."

She added: "He likes to write when he's under great pressure, to express his thoughts and feelings. Sometimes when there happens to be no paper available, Simon uses paper napkin as an alternative. He writes for himself and for others. In early years, he tended to have a temper under pressure. He's more patient now, the tasks he assigns can be done by tomorrow instead of within taday." She says in a joking tone before becoming more serious, "But he really cares about his employees. There is caring in many things he does. It is always about benevolence."

Cherry continued: "When employees or their family member fall ill, Simon would support them personally in additional to the amount provided by the company: he doesn't want to break the company's rules while he really wants to help. He used to quote from the *Book of Songs*, "Those who know me well may ask what is bothering me; yet strangers only ask me what I want." What the company has is earned, penny by penny, by its hard-working team. The company will never forget what every single one in the big family has contributed. It's rare to find a boss like him."

"It's the same when he's doing business. Sometimes it's obvious that he is duped by others, but he doesn't mind while we are really worried. Many things he does are not provoked by self-interest. Instead, he seeks to contribute benevolently to the betterment of society. For Simon, it is always about benevolence."

Constant in the way of living and dynamic in decision-making; a big picture reader and small detail observer; rational and perceptual; gentle and impetuous; persistent and flexible; modest and confident, Simon manages to combine so many binary characteristics harmoniously in doing business. As portrayed in a Chinese poem, "Sitting by a raging river, my heart is at peace, as the cloud above my head, at ease." It's his simple and unchanged heart that makes him a successful person.

CHAPTER

# IX

# RATIONALITY

/

## UNDERSTANDING THE TAO

# Truth-Seeking, Fact-Pursuing and Value-Persisting

The bookshelf in Simon's office by the seaside, takes up an entire wall, where different kinds of books can be found, ranging from history, culture, art, sinology, Chinese medicine, business management to biography. At the side, there is a long bench where several piles of books are scattered. Simon likes to write down a line that he finds inspirational. Whenever he reads a good article, he will save it for continued study. He is a diligent learner. Having not much opportunity to read at young age, as Simon always said, he now treasures every chance of reading and learning.

The office was composed of two rooms separated by a sliding glass partition, this unique design feature giving the office the appearance, from a distance, of spaciousness and brightness. Simon selected every detail of the design, including the furnishings which he found from different places. These items show their owner's taste. He was also attentive to matching the items as the season changes. When autumn arrived, Simon replaced the ink painting at the entrance, which illuminates a mountain with water shimmering from a distance under the sun's glow. Two oil paintings were, as well, affixed to each wall of the office: the first showed a colourful abstract surfaced with various intensities of colour. The other pictured pink azaleas on a green mountain — a simple yet appealing scene.

"Not all these paintings are from great masters. I would buy a painting as long as the artistic conception is good, just like this one." Simon said, pointing to the oil painting on the left wall. "The painter is French. She majored in engineering at school and worked as a bank manager, didn't start painting until 40 or so, but is very talented."

Those who have visited Simon's office marveled at its distinctive design. A Hong Kong friend in real estate once visited Simon's office. When the friend was about to leave, he told Simon would like to return if only restudy the design. Had seen many new offices with many different designs, he said Simon's office was the first that inspired him with the desire to revisit. "The office is not only traditional and elegant but also stylish and modern, a very good combination."

A colleague who participated in the decoration project recalled that Simon

❖ *Simon's office*

was very sensitive to space and colour. "For example, he would ask us to make the ceiling higher, so the office could look more spacious. He also had a holistic view of the design so that the colour of the wall contrasted effectively the tone and texture of the furniture. From a brief survey of a design sketch, Simon could immediately spot the appropriate distance between two low cabinets. It was astonishing. The visual effect was totally different once adjusted the space between the furniture pieces."

In considering design, Simon attaches great importance to the concept of "spheres".

"The dialectic concepts, high and low, far and near, interior and exterior, square and round, are all contained in the term, spheres."

Simon sees spheres as speaking to both the physical environment and the psychological mood. The two are mutually reinforcing. The essence of design is to have these two "spheres" matched. "In the same vein, the soul of a painting is also the artistic conception of scenery and feeling," Simon explained.

Simon's connection with art began with a painting that his father gave to him. When Suen Shing's company had been closed down, he asked Simon to take this painting by the esteemed Chinese painting master Mr. Xu Beihong. "My father said this painting was as valuable as his entire fortune," Simon said. "He insisted on giving this painting to me, asking me to pass the work from generation to generation as a family treasure."

His father had wanted to compensate Simon, his elder son, for having taken good care of him and for paying the monthly rent for his office. At first, Simon refused to accept the gift. It was his filial duty after all, but his father insisted on making the present. So, Simon enclosed the painting in a piece of wrapping paper and kept the item for more than a decade. "I didn't know anything about the work until I met Dr. Kwok Ho Mun of the Wan Fung Art Gallery, who heard me say that I had a painting by Xu Beihong and wanted to see the object. So, I took it out for him to have a look." When Simon carefully opened the paper wrapped around the painting, layer by layer, and displayed the scroll in its entirety, Dr. Kwok took one look, then another. A look of bewilderment surfaced on his face. "Simon, this is a print, don't you see?"

Simon laughed, "Of course, I didn't dare tell my father the truth. The painting was a fake. I gave the print to my daughter afterwards, who was studying in the United States, to decorate her apartment in New York. It was ten years later that I had my first authentic painting by Xu Beihong."

"But even at that time, I was a layman who knew nothing about art and was thinking in business terms about what to buy and where to buy. I chose to go Sotheby's because I believed in its reputation. With my first bid, I aimed for the most expensive item, believing that price reflected value. That year, Sotheby's put up hundreds of paintings for auction. The three most expensive works were by Zhang Daqian, Xu Beihong and Wu Guanzhong. In fact, they were all great artists, and the three paintings were all masterpieces, but I chose Xu Beihong's because there was a personal story there."

During the SARS outbreak in 2003, the Hong Kong economy was in the doldrums and the art market was lackluster. However, Xu Beihong's *Eight Horses* became the target for relatively active bidding, Simon competing against a number of potential buyers. He was so determined to win after each competitor offered a bid that he raised his price. The contest drew the attention of connoisseurs in the collecting world, such as Linus Cheung Wing Lam, a major collector and former director of Sotheby's in 2016. "He had never seen me at an auction before and didn't know me. He thought I was some kind of mysterious nouveau riche," Simon said with a self-deprecating smile. In the end, he won the auction and gained his first real painting by Xu Beihong. He and Linus became good friends and often went to auction previews together. Since then, the biannual Sotheby's auctions have become a regular part of Simon's schedule.

"I don't speculate on property. I just like to collect paintings. I have a rough budget every year and set a bottom line before the bidding, so I wouldn't become impulsive and get carried away." The large bookshelves lining the wall of his office contain an entire shelf of auction catalogues of paintings and drawings from Sotheby's and Christie's, arranged by year and in complete order. "In the bidding process, it is important to remain calm, accurate and precise; to recognize when you need to be patient and when you should be ruthless. At the same time, it also depends on the opponents. It shouldn't be forced; it is about destiny."

Whether doing business or collecting art, Simon adopts a consistent approach basing his perceptual understanding of the situation and analyzing the circumstance in a step-by-step and highly rational manner.

He retains fondness for paintings by modern master artists, such as Fu Baoshi, Qi Baishi, Xu Beihong, Zhang Daqian, and Huang Binhong. "Their works are not constrained by time and space: Chinese people from both the south and the north can buy them; Asian people can buy them; and Westerners can also buy them. They were popular fifty years ago and will still be popular fifty years from now. Such works have commercial value and are worthwhile collecting."

"He used to buy the painting on the cover of the auction catalogue," says Raymond Leung.

Raymond would often accompany Simon to art exhibitions and auctions. He explained his friend's reasoning that the cover work, normally best of the best, which served as a magnet to draw customers in, must have a strong selling point to be recommended by an auction house.

Simon started his business as a salesperson and understands the psychology behind effective marketing very well; he applies a business mindset to identify works that fit with his unique vision. The more works he has seen over time, the more he is able to ferret out the rationale for a particular work that has become featured on the cover. He has generally been able to grasp the market and connoisseur's criteria of judgment, so he could select his favourites from hundreds of auction items every season.

"To judge whether a Chinese painting is worth collecting, you have to see whether the painting contains a poem, and whether there is a story behind the painting."

*Simon was appreciating Shi Tao's "Lotus Society" with his friends. From left: Chan Lee Chung, Hung Hoi, Linus Cheung, and Simon*

There is a glint of quick-witted pleasure in his eyes, "It's actually a test of my own judgement. Collecting art is also an investment, and I enjoy the art as well as the pleasure of making money."

Nevertheless, even though he tends to consider art as, in part, an investment, Simon has almost always bought but rarely sold work. And he treasures his favourite pieces. When he is in a particularly good mood, would invite friends to get together and enjoy the paintings. Simon is a truly collector rather than a speculator who has his own systematic approach, not only considering a painting's value but also its meaning in the context of his developing collection.

"When looking at paintings, one should first consider the history of Chinese art and start from a broad perspective. When it comes to buying a specific painting, I do my homework, finding relevant books to understand the artist's life story, the year of creation, and the background of a particular period. It's not just about appreciating the painting; it's also about understanding the person." Simon follows the strategy of "letting go" of preconception before "taking in" the meaning of an artistic object. "By adopting the two-pronged principle, one can be open-minded in appreciating all kinds of work."

Simon likes to visit art museums on his overseas business trips. "Many famous masters have studied abroad, and the influence of Western techniques has brought out a different flavor of Oriental expression. We need to see the similarities as well as the differences in order to better understand the integration of Western technique into an Oriental style and thus, to understand the significance of each artist's innovation: we can, as a result, gather in greater depth the uniqueness and value of a painting."

Simon summarizes the essence of his buying strategy into the following key words: refinement, authenticity, newness and depth:

"Refinement: Every modern master artist has a peak period, and the works produced during this period are the most valued. An example would be Zhang Daqian's watercolour splash paintings created during his 60s.

"Authenticity: To distinguish the genuine from the fake, the source of the work must be well documented, and the story behind the work must be able to stand up to examination.

"Newness: The work should be well maintained, free from moisture and appear in quality to be new.

"Depth: to become a collector, one must hold true through the test of time, then enjoy the collection, allowing its profound meanings to seep onto the spiritual level."

❖ *Photo taken at the Musée d'Art Moderne, Paris*

❖ *Luciano Benetton (left), co-founder of the Italian brand Benetton, met Simon in business, and they both love art. The photo was taken at Luciano's collection in Italy.*

Traditional Chinese culture has a subtlety rooted in its interconnectedness among Chinese medicine, martial arts as well as calligraphy and painting. Simon has a broad-ranging background that trains him to decipher the intricacies commensurate with developing a collection.

His practice of Ba-gua and past reading of Chinese medicine classics have permitted Simon a special insight into the artistic features of Chinese landscape painting. Simon realizes that elements apparently in opposition, in fact, share a dependency. Thus, he sees the bold and supple, the solid and the void, the static and dynamic, the yin and the yang as two sides of the same coin. The recognition of these synthesized pairings sparks for Simon a deep understanding of art as a form. "In fact," Simon remarks, "no matter the artist, his work is inseparable from the features embodying the Chinese philosophy."

From a layman to a lover, Simon's philosophy of art collection is quite discerning and sophisticated. It is built on the subtle recognition of the dialectic in bringing apparently contradicting aspects of an aesthetic object.

"Here is the 'guiding principle'. The art of calligraphy and painting becomes encapsulated in the brush's movement and in the ink lines that the spirit of mountains and waters on the paper. The essence of culture is reflected in the harmonious unity of spirit and matter, and in the unity of subjectivity with the objective representation of reality: a fused expression epitomizing harmony and balance in spirit and form."

"The grasp of the essential understanding then enables the synthesis of the interior scene with the outside world evoked by the imagery, so that one can experience both the story of the painting and its poetic meanings from the artist's delicate application of brush and ink. This is the 'substance'. A masterpiece should be able to encapsulate the essence and embody the ingenious talent of the artist. For example, Zhang Daqian's watercolour splash is unprecedented; just as Xu Beihong notes, 'Zhang Daqian is a once-in-five-hundred-years genius', whose reputation is well deserved."

Simon's eyes sparkled with excitement while he talked eloquently. On the eight-foot wall behind him was a thick and lavishly green landscape painting by Zhang Daqian, like a poem said, "Mountains upon mountains, mountains far, sky high, misty water cold, and the longing like frosty maple leaves grow."

The gift of his father's painting foretold Simon's predestined affinity for the art of painting and calligraphy. As he entered into the business world, Simon had a series of coincidental encounters with the world of art, and befriended many artists, including masters like Wu Guanzhong and Fan Zeng.

One of the first artists Simon met was Cui Ruzhuo, who returned to Beijing after a decade of living overseas. Cui studied painting under Li Kuchan and was skilled in painting flowers and birds, but at that time, he was not very famous yet. Whenever Simon went to Beijing, he would invite Cui to meet him for a drink; after drinking, they would do palm presses, each performing a hundred reps. The competition in drinking and sports served as a foundation for a friendship between these two gentlemen, one a painter and the other an entrepreneur.

"It was a pleasure to visit Fan Zeng's home, watch him write and paint, have a home-cooked meal and a pleasant talk." Simon recounts his first visit to Fan Zeng in 2006. Later, when Fan Zeng's exhibition opened in Jiangxi, he sent an invitation to Hong Kong, inviting Simon to attend the ceremony. "But it was Eric's first day

❖ *Simon (left) and Cui Ruzhuo*

❖ *In Wu Guanzhong's apartment in Beijing*

❖ *A visit to Fan Zeng, during which Fan improvised with his calligraphy. On the left was Simon's friend Raymond Leung.*

of secondary school, and as a father I must be there with him, so I couldn't visit the exhibition." Simon showed a little bit of regret. It was clear that he treasured his friendship with each artist.

Simon has also been close to many local artists in Hong Kong, including He Baili, the fourth generation of the Lingnan School, and Hung Hoi. The latter studied under the famous master Yang Shanshen.

"There are several miniature landscape paintings by He Baili on the wall of our office. His ink landscape paintings exhibit the development of the artist's individual style in a traditional art form; the ink lines are so fluid that critics have called them 'He's landscapes'. I have learned art knowledge from him, which has inspired my future collection. To some extent, he can also be called my art enlightenment teacher."

Hung Hoi, a local painter, became Simon's friend and teacher. Hung grew up in the Mainland and had studied under his own father's tutelage since childhood. Once arriving in Hong Kong, he worked under the mentorship of Yang Shanshen, a well-known artist. Hung is skilled at sketching landscape, applying fine brush strokes. His compositions have majestic settings: the lines are dense and smooth, expressing a natural style and spirit. It is said that he draws as if hills and valleys are envisioned.

Hung, a gentleman with graceful manners, also teaches in a university. Simon said that he had gained a great deal of knowledge about art from Hung. The two friends would often invite each other to see exhibitions together.

"Simon has a good eye and usually can, with a quick glance, appraise the finer points of a painting. He has his own taste, and his standards are so high that sometimes he finds nothing even after surveying an entire collection. He is also very careful in choosing paintings and will do a great deal of research to study a work. But when he sees a particularly good one, he is bold enough to bid for the item, and during an auction, he can become quite aggressive."

"He is a courageous and wise man," said Hung. "Due to his accurate judgments and his proactive bidding, Simon has managed to collect many representative works of art." Hung smiled and continued, "And he really loves art and would be so happy when bought a good piece, even sending me a message to show the work. He is very easy-going, friendly and never imposes his point of view. We used to go to Beijing to see exhibitions together and have a good time."

As a businessman, Simon has made many friends in the arts world, including Cheng Shifa, a master of ink and wash, and Lin Jin, a calligrapher and seal carver.

Han Tianheng, the current Vice President of the Xiling Seal Engraver's Society and a great calligrapher and seal carver, Yu Zhongbao, an oil painter in Shanghai, and Xie Chunyan, an art critic, also regard Simon as a good friend.

His younger brother, Suen Siu Wing said of his brother: "There is definitely something about his personality. From childhood to adulthood, he has made many friends from all walks of life."

"When I was young, I was most impressed by the fact that some people from the city with a higher level of literacy, or overseas Chinese who returned to their villages to become teachers, or even senior villagers, all became good friends with my brother. It was really rare and uncommon, given his age, education and our family background. This was because of his unique temperament and character," said Suen Siu Wing, who admire Simon's charismatic qualities a lot.

"There are many reasons why a young man less educated may be recognized and affirmed by so many educated, qualified and senior people. Simon has an extraordinary respect for knowledge, for culture, and for knowledgeable and cultured people. The most typical example is his friendship with Professor Jao," Suen Siu Wing says after a pause.

Professor Jao Tsung-I was a great scholar of his generation and a dual master of both academe and art, who has made vital contribution to nearly every field of traditional Chinese culture.

In 2012, when the Hong Kong Baptist University awarded honorary doctorates to Simon and Professor Jao, the newspaper headline was "The Master and Disciple Honoured Together."

Simon considered it inappropriate. "Professor Jao is so erudite that I can't even understand his books, can't be called his disciple. I call Professor Jao teacher out of respect. I've had the privilege of being around him for so many years and am most happy if I can learn a thing or two from his conversations," Simon said.

One was an internationally renowned scholar, and the other was an entrepreneur committed to innovation. Their fields resembled parallel lines that never intersected, yet they did. It might be hard from the outside to understand their relationship.

Simon didn't care if he was distrusted and never defended. One day, a high school classmate of his who had since become a noted businessman, being familiar with many celebrities in the city, invited Simon to lunch and told him that Professor Jao had been invited and promised to attend.

Simon said, "I couldn't be absent if Professor Jao was there."

❖ *Simon(left) and He Baili*

❖ *Simon with Lin Jin (right) and Han Tianheng (middle)*

❖ *Simon with Mr. Xu Qinping, son of Master Xu Beihong*

❖ *Simon with Yu Zhongbao (left) and Xie Chunyan (right)*

The lunch was at an old-fashioned Cantonese restaurant on Hong Kong Island, tucked away on the corner of a bustling street, which one could easily miss. The interior decoration was simple, but the food was outstanding. The restaurant didn't use advertising but had, nevertheless, become a popular hotspot. The guests were famous in their own right. In the middle of the lunch, one of the guests half-jokingly said to Simon, "I heard that you always peeked at the homework of the classmates sitting next to you in elementary school?"

Obviously, the intention was to embarrass Simon. But Simon smiled and replied, "I can't recall much from my childhood."

Professor Jao, who hadn't said much, commented: "Simon is a man of wisdom." The room became immediately quiet. "Professor Jao understood me," added Simon with a sense of gratefulness, "Having knowledge is not the same as being knowledgeable; being knowledgeable is not the same as being talented and being talented is not the same as being wise."

As for the friendship between Simon and Professor Jao despite their vast difference in age and the distinction in their professions, Suen Shing, Simon's father, once asked Professor Jao seriously, "Of all the people you know, why do you favour Simon?"

Professor Jao's eyes shone brightly and replied, "It's all about fate."

The story of this fate began on an occasion: *"A Symphony of Academics and Art: Exhibition of Jao's Artworks"* held by the Soka Gakkai International of Hong Kong.

"I remember when I entered the exhibition, there was an eight-foot, four-fold lotus painting with magnificent brushwork and elegant poetic charm."

The first time he saw Professor Jao's works, they made a deep impression. Mr. Lam, a graduate from the Department of Chinese in the University of Hong Kong and an old acquaintance of Professor Jao, accompanied Simon at the exhibition. One weekend afterwards, Mr. Lam took Simon to visit Professor Jao.

"The first time I shook hands with Professor Jao, his hand felt so soft but also strong, not at all those of a man in his eighties. Hardly to imagine!" Simon talked about his first encounter with Professor Jao.

Simon expressed his amazement, "Such a remarkable scholar lives in a mere 700-square-foot house. There are books everywhere. I was particularly touched by Professor Jao's comment that the space is small, yet the world of learning is big. This is what it means a quiet mind for learning and a room full of wisdom. When I left, Professor Jao personally saw me off to the door where he treated me, a mere small potato, with an admirable degree of modesty and courtesy. His erudition and temperament greatly impressed me."

At that time, Professor Jao was in the process of compiling his life's work into a collection entitled, *Collected Works of Jao Tsung-I*, covering a range of subjects, including Confucianism, Daoism, Buddhism, poetry, lyrics, literature, history, bibliography, archaeology, Dunhuang studies, phonetics, calligraphy, painting and oracle bone inscriptions. "He insisted on proofreading every word of the 14-million-word collection himself." Professor Jao was so exhausted that he suffered a slight stroke. When Simon learned of the incident, he offered to arrange a place for Professor Jao to rest and recuperate. With the consent of his family, he brought Professor Jao to Dongguan to recover in a quiet place with beautiful scenery under the Lianhua Mountain. For several months, he also took good care of Professor Jao until the master recovered completely.

"Simon was really nice. When he saw Professor Jao's condition, he was willing to offer help and arranged for him to live in Chang'an for a while. The place was very nice, and he was with Professor Jao each day; that was very considerate of him. Not only him, but also his colleagues in the company were very nice, often visiting Professor Jao and taking care of him." Professor Jao's younger daughter, Ms.

❖ *Professor Jao and Simon at the foot of Lianhua Mountain*

Angeline Yiu, recalled. Ms. Yiu had a teaching career in her early years and later, attended full-time to the care of Professor Jao, handling both his personal and public affairs.

During Professor Jao's convalescence, Simon spent time with him and became more and more impressed by his scholarship and personality. Professor Jao also saw Simon's true character of sincerity and respect for knowledge. After returning to Hong Kong, Professor Jao and his family hosted a dinner to thank Simon and wrote a poem, praising his sincerity and righteousness.

Professor Jao had seen many people and was a perceptive judge of character. He was right about Simon. Since then, Simon remained unceasing in his care for the Professor.

In 2004, when Professor Jao went to Chang'an again to recuperate, Simon took care of the arrangements. When the Professor went to Beijing, Shanghai, Guangzhou and other places to attend the major exhibitions, Simon was the one who planned and coordinated, even attending to the details about transportation and accommodation. When he invited the Professor to dine, he would wait at the drop-off location, always remembering to ask for one more cushion for the chair where the Professor sat. Simon would always send the Professor to the car first after the meal, and he would never leave until the Professor's car started. On a business trip to London, Simon picked out a scarf for the Professor at a British store specializing in wool, comparing the items, caring more of the Professor than he did for himself.

Finally, Simon chose a classic one with checkered stripes, which the Professor wore with great spirit. When Professor Jao was almost 100 years old, his hearing was not as good as it used to be. Once Simon had lunch with someone from the technology industry and inadvertently learned about the latest hearing aids. He took the information to heart and immediately ordered one set for Professor Jao. Ms. Angeline Yiu said the Professor could hear more clearly after wearing the device. It facilitated communication, and the Professor was willing to speak a few words from time to time. One day he was so happy that he even recited some passages from *Travels of Fah-Hian and Sung-Yun, Buddhist pilgrims, from China to India*, and he got every word right. Simon, who was looking at him, laughed and clapped his hands. He was happy from the bottom of his heart. Every year on the first day of the Chinese New Year, Simon would go to pay his respects to the Professor; on his birthday, Simon would celebrate with him. As with an event of significance in these trivialities, he had always respected and sincerely cared for the Professor.

❖ *From left: Ms. Angeline Yiu, Professor Jao and Simon at an event (2002)*

❖ *Simon and Professor Jao had lunch together; they had a good time looking at the photos on Simon's cell phone (2012).*

"My brother's relationship with Professor Jao was very special. He did not receive much in the form of formal education and had, nevertheless, met one of the most respected figures in modern times, a great scholar of both the East and the West. The fact that two individuals from completely different fields were able to maintain such a relationship, I think, was born out of his deep respect for teachers. That attitude of respect and consideration deeply moved both Professor Jao and his family. Not just during the few months when Professor Jao was recovering from his illness, but since he developed a relationship with Professor Jao, he had been quietly giving and doing a lot of things." Suen Siu Wing, who has been by his side all these years, could see very clearly what his brother had done.

In 2003, Simon accompanied Professor Jao — who had recently recovered from an illness — and his family to Chaozhou to visit his former home, Tian Xiao House of Chun Yuan. Chun Yuan was originally a private residence of the Jao family, built by his father, Jao E. The two-story Tian Xiao House was elegantly designed. The doors and windows were made of imported stained glass from the West and were given a distinctive and elegant pattern that was bright and dazzling. In 2019, the Hong Kong Heritage Museum held an exhibition entitled "The Story of Jao Tsung-I" and reproduced one such window to display in the exhibition hall. The Tian Xiao House constitutes the library of Jao's family, and the place where Professor Jao spent most of his childhood. The Professor was very excited to revisit the place. Simon sat on a wicker chair next to the Professor who recounted his past. Lotus flowers thrived in the courtyard behind the pair.

"Professor Jao came from an educated generation. His father was also a learned scholar. Tian Xiao House was the largest library in eastern Guangdong at that time. Professor Jao studied in that house from his childhood on. He did not go to college but received a thorough education at his family's home. His academic skills and achievements were unmatched among his peers."

"Professor Jao always told me that one should be able to tolerate loneliness and solitude when studying. This kind of 'pure mind' has never changed since he was a child. He woke up every morning at 4 AM or 5 AM, read and wrote while expressing a full devotion to his studies." Simon was respectful and full of affection when speaking about the Professor. "He spent his whole life teaching and performing scholarship. After retirement, he still kept on exploring, and developing new reservoirs of knowledge. He has been a pioneer in many different fields. He was the first scholar to translate ancient Chinese bamboo and silk manuscripts discovered in

❖ *Professor Jao and Simon at Tian Xiao House*

❖ *Simon likes the unique design and shape of the stained-glass windows of Tian Xiao House.*

❖ *From right: Simon, Ms. Angeline Yiu, Professor Jao, Mr. Lam, and Dr. Thomas Tang*

Changsha, Hunan Province."

Simon referred to the bamboo and silk manuscripts of the Chu State in ancient China, unearthed in 1953 from the Chu Tomb of the Yangtian Lake in Changsha, Hunan Province. Subsequent to that excavation, Professor Jao published a number of studies, including *Zhanguo Chujian Jianzheng*, *Changsha Chutu Zhanguo Cujian Chushi*, *Chujian Xuji*, the very first studies on the Chu bamboo and silk manuscripts widely acknowledged in the academia.

"Professor Jao had a famous saying: 'Scholars don't like the hustle and bustle.'" Simon looked at the 20-volume *Collected Works of Jao Tsung-I* neatly placed in his bookshelf and said, "But he was not studying alone on an isolated island, yet having numerous exchanges with other scholars."

In 1970, Professor Jao debated with Professor Hsu Yun Tsiao, published an article to prove that his contestant had made an error in his analysis of the book, *Imperial Recordings of the Taiping Era*, while Professor Hsu admitted his error shortly after and replied in an article. This debate became a well-known story in scholarly circles. Stories, such as this monumental contest, had a deep impact on Simon.

"Professor Jao was most fond of Su Dongpo, one of China's most famous poets and scholars. Su Dongpo said, 'Learning is like accumulating wealth and there should be nothing left.' What he meant was that a scholar should take the same approach as the wealthy, attempting to accumulate as much knowledge as possible. There is a book called *How the Steel Was Tempered,* which shows how a wound tempered a soldier into a great warrior. If you look at Professor Jao, you will know similarly how a great scholar owned the name.

"I couldn't understand many of Professor Jao's books, but he would tell me stories and make allusions to scholarship. Sometimes when I didn't quite understand the meaning, he would take a piece of paper and write out some lines down for me. He would also talk about it during meals, and when he got excited and couldn't find paper, he would drop a note on a napkin." Simon had carefully collected each napkin, storing away the wisdom.

When he found out that the Taiwanese publisher, which owned the rights to the traditional Chinese edition of the *Collected Works of Jao Tsung-I*, had only printed 300 copies of the book and had given only 20 copies to Professor Jao, Simon felt a great pity.

"Such a good collection of books should be printed in greater numbers, and every library and research institution should have them." With the support and cooperation of many parties and his careful arrangements, following the publication of the traditional Chinese edition in 2003, the simplified Chinese edition of the *Collected Works of Jao Tsung-I* was published in late 2008. This edition included fourteen volumes and 20 books, collecting all of Professor Jao's major works and covering almost all areas of Chinese studies. Simon ordered 100 sets and gifted them to universities and research institutions.

He said, "I don't understand much about the content of this collection, but I know a lot about its value."

Professor Jao's academic works are so precise, specialized, and profound that they are not easily accessible even to other scholars. But there is one of the Professor's books that Simon often reads: his 1999 collection of poems, lyrics, verses and prose entitled, *Qing Hui Ji*.

"Every now and then, before going to bed, I flip through *Qing Hui Ji*. This book covers Confucianism, Taoism and Buddhism, and each time I've gained a different experience. For example, page 248, on the rhyme of *Qing Tian Ge*, there is a paragraph about the poem of Xu Wei of the Ming Dynasty: 'More ink or less ink are all barriers; the ink is full and the brush is fast to represent the universe ...' Every time I read it, I feel enlightened and have an epiphany. My understanding is that calligraphy should be combined with the Tao through the unity of mind and spirit so as to reach a level where all is in harmony with the natural world."

Professor Jao was so impressed by Simon's heartfelt realization that the Professor wrote Xu Wei's poem with a specially made pen the calligraphy, Simon hung the calligraphy in his bedroom where he could each day appreciate the work.

"Not only was Professor Jao a profound scholar, but he was also an accomplished artist — something not many know."

Professor Jao has written extensively on painting and calligraphy artists throughout the ages, from the "Four Wangs", who were Chinese landscape painting artists of the Yuan Dynasty, the "Five Masters" seal carving, an art form from the Ming Dynasty, to the "Four Masters" of philology of the Qing Dynasty. The nineteenth book of the *Collected Works of Jao Tsung-I* is devoted to art and includes a collection of Professor Jao's essays on the history of Chinese painting, *Hua Ning*.

"I have a work by Shi Tao, which I have collected for many years. Zhu Qizhan, a master painter, had been its previous collector." Simon bought it at a high cost, said, "If that price for that work had been used to buy land, I could have bought several plots."

Later, he showed the painting to Professor Jao. As an authority on the "Four Monks" — Ba Da Shan Ren, Shi Xi, Shi Tao and Hong Ren, who were famous painters from the early Qing dynasty — the Professor was very fond of the painting and studied the work before requesting to see it again. The second time he saw the work, Professor Jao examined the work for two weeks. Considering it in depth, he remarked that it was the best piece of Shi Tao's works he had ever seen, even wrote a 200-word inscription in honour of the painting.

Professor Jao said to Simon, "Your good fortune, my eyes' feast."

Simon's box room was named "Yi Tao Ju", standing for the prominent place the work by Shi Tao held in Simon's collection.

Professor Jao's paintings and calligraphy embody his own unique style. "The fact that his painting of lotus flowers is called 'Jao Lotus' and his calligraphy is called 'Jao Style' bespeaks everything," Simon expressed with admiration.

"From my experience, the mood and technique of a painting should be complementary. Professor Jao had a rather interesting way of describing his paintings as the 'trinity' of CPP: C stands for calligraphy, and the two Ps stand for

❖ *From left: Dr. Thomas Tang, Simon, Ms. Angeline Yiu, and Professor Jao, admiring Shi Tao's work "Tiaoxi Shiyi Tu" together*

painting and poetry. The acronym exhibits the completeness and comprehensiveness of the Professor's masterly approach." Simon admires Professor Jao for his ability to make profound theories accessible to the general public.

"He painted lotus flowers with large strokes of ink completed in a single breath. For instance, his lotus stalks were painted from top to bottom in a single stroke. Professor Jao told me that even if the lotus is a few feet in height, he used his breath to control the ink and brush. Hearing him say that, I understood why the four-fold screen of lotus flowers I had seen on my first visit was so robust and powerful. This is also a skill that he had practiced over the years in writing large characters of calligraphy. He wrote smoothly and in one breath, even a large character of three feet in height. The inscriptions he dedicated to the 'Wisdom Path' at the foot of Lantau Peak in Hong Kong in 2002 were precisely that size. They were carved on Chinese rosewood pillars."

"Seeing the economic downturn in Hong Kong and the widespread of pessimism at that time, Professor Jao wrote the *Heart Sutra of Prajna Paramita* after recalling the cliff carvings he had seen in Mount Tai. Later, the government built the world's largest outdoor wood inscription of Buddhist sutra, 'the Wisdom Path', at the eastern foot of Lantau Peak, in the hope of encouraging and comforting its citizens. The inscription shows Professor Jao's caring and compassionate heart for humanity."

Simon recalled, "Professor Jao told me that he had been to Hong Kong when he was a teenager, to help collect debts for his family's business. When he saw that those in debt lived a tough life, he did not want to force them to pay and returned with empty hands." At that time, Jao's family was the richest in Chaozhou, running several money shops, old-style Chinese private banks. "He has been compassionate since he was a child."

At the entrance of the Jao Tsung-I Petite École at the University of Hong Kong, there are two wooden pillars carved with Chinese words "Ci Bei" meaning "compassion" and "Xi She" that together equates with "joyful giving." Professor Jao drafted the calligraphy in his own inimitable style.

The Petite École was founded with Professor Jao's unconditional support, who donated over 34,000 rare books and journals, collections of articles, and over 180 pieces of his calligraphy, paintings, and artworks to the University of Hong Kong. Professor Jao created the title, "Jao Tsung-I Petite École". Petite École translates as small school, and "Professor Jao described himself as a 'small student' and could not give this place a big name. He was being humble and humourous," Simon

❖ *In January 2005, as the Founding President of the Jao Tsung-I Petite École Fan Club, Simon (first from right) accompanied Professor Jao Tsung-I (second from left) to the foundation laying ceremony of The Wisdom Path, together with the other two Founding Presidents, Mr. Chan Wai Nam (first from left) and Professor Lee Chack Fan (second from right).*

remembered.

In August 2010, Professor Jao went to Beijing to visit Professor Ji Xianlin, who was hospitalized at No. 303 Hospital. "The newspapers described it as a historic meeting of the century between two equally famous scholars: Jao based in South China and Ji in North China. Professor Jao said there was no such thing as a meeting of the century. They were just two old fellows trying to flatter each other." Simon, who accompanied Professor Jao, recalled, "Since he was often referred to as a historian, an archaeologist, a literary scholar, scholar of classics studies, calligrapher and painter, Professor Jao joked that he was impossible to be classified and he belonged nowhere."

Soon after Simon became acquainted with Professor Jao, he once, at great expense, hired a musician to perform guqin, a seven-stringed plucked instrument in ancient China, to celebrate Professor Jao's birthday. When the musician arrived, he put down his instrument after seeing Professor Jao and gave a big salute before turning around to leave the room, leaving Simon bewildered. "I went after the musician to ask him why he left. He told me that as Professor Jao was a great master of guqin, he wouldn't display his slight skills before an expert."

"Professor Jao never bragged. The breadth and profundity of his scholarship matched his great personality. I have never heard him gossip about others, and he never fussed over money. He didn't even have a bank account. He lived a very simple life. Later, he often had to attend events, hence a few suits were prepared after discussing with Angeline, his daughter."

In 2012, Simon accompanied Professor Jao, 97 years old at the time, to visit Kyoto University in Japan. Professor Jao was pleased to see his former students who translated his works. He was so happy that he wanted to compose a poem by caligraphy. On the spur of that moment, he could not find any xuanzhi, a type of high-quality paper made for traditional Chinese calligraphy and painting. So he just inscribed on a piece of standard paper "*Ting Shu Seng Jun Tan Qin*", which translates loosely as "On Hearing Jun, the Buddhist Monk from Shu Playing His Guqin". The work was written by Li Bai, an immensely important poet from the Tang Dynasty period:

*The monk from Shu, holding his guqin-case green,*
*Comes westwards down Mount Emei serene.*
*With a stroke of his hand, he plays for me —*
*From a thousand valleys — waves of pine tree.*
*My heart is cleansed in the guqin's brook running.*
*The clear frosty bell's echoes are stunning.*
*I see not the blue mountains darkening,*
*Nor the layered autumn clouds, when I'm harkening.*

In Simon's opinion, this story showed how deep Professor Jao's passion was for poetry and how he valued his friendship with others.

When Professor Jao visited Kyoto University previously, he stayed in a temple in the suburb. "Professor Jao told me he might have been a monk in a past life, and I believed that. His experience in this life was much similar to a monk's. He first acted upon his belief in Confucianism, later found truth in Taoism, and eventually with Zen found his inner peace." After having known Professor Jao for nearly 20 years, Simon had grasped the teacher's poetic heart and the Zen mind.

"It is essential for people to settle properly in this world. Maintaining a free mind is a state of being. When Professor Jao mentioned settling down, he was talking about a stage that he had reached where he had become free of worries," Simon continued. "Professor Jao said that if people understood the wisdom of Buddha, they would be indifferent to the temptations of the material world. They would live in the present moment. Any losses and gains in life were destined. Not caring about success or failure, honour or disgrace, and gain or loss, and letting go of obsessive desires, they would set their minds free of worry and become relaxed. They would feel content with their life and have a joyful spirit. They would live without regret. That is true wisdom."

"I was there when Professor Jao passed away. I saw his face unchanged, his heart unhindered. His death was peaceful. Not all can have such blessings in their life." Simon lowered his voice.

During his trip to Kyoto, Professor Jao was not as strong on his feet as before. Simon pushed his wheelchair, and the reporters, who accompanied the visit, described Simon as the Professor's "navigator". However, in the journey of life, Professor Jao was Simon's navigator. Inspired and influenced by Professor Jao, Simon learned much more than about the intricacies of traditional art and culture but also grew to understand deeply the most important philosophies of compassion and joyful giving. He gradually shoulders on the undertakings to spread and promote traditional Chinese culture and has made the endeavour his responsibility.

Simon discussed Professor Jao's "One Tree and One Road."

In 2015, at Professor Jao's 100th birthday dinner, a special video was made to show the fruits of his extensive work in numerous and diverse fields. It illuminated a tree that, during the video, grew, flowering before its audience. Professor Jao's daughter, Angeline Yiu , had the creative inspiration for the stunning theme.

"There is one tree that is important to Professor Jao," Ms. Yiu said. She referred to his book, *The Tree of Chinese Characters: Symbols, Proto-Writing and Ideograms*. In this book, Professor Jao explores the structure and evolution of Chinese characters in the primitive era and how they have continued to be used over the millennia, their shapes remaining unchanged. Ms. Yiu said, "Professor Jao was a master of many different languages, including English, French, Japanese, German, Indian, and Iraqi. He also knew Sanskrit and Ancient Babylonian cuneiform characters. He once told me that it was remarkable that Chinese characters had remained unchanged for thousands of years in a country as large as China, and that it was only after learning so many different languages that he had realized the greatness of the Chinese characters."

These words touched Simon deeply.

"He truly understood the value and meaning of traditional Chinese culture," Simon said.

"Professor Jao once said, during the westernization movement in the 1860s to 1890s, many people had lost their confidence in traditional culture because of the impact from foreign cultures, and intellectuals also turned away their own culture in pursuit of foreign culture. This is a case of trying to 'know one's opponent' without 'knowing oneself.' We all must reflect upon this. He said on many occasions that the twenty-first century would be the era of 'East to West' where Chinese culture would be introduced to the western world and that China would embark on a Renaissance. He held an outlook on traditional culture based on his profound understand of history and his sense of responsibility to disseminate traditional culture. Professor Jao was most fond of painting lotus flowers. The lotus flower expresses a willingness to bear the burden and take responsibility. Thus, the flower reflects Professor Jao's lifelong attitude towards his studies."

Simon becomes emotional and full of memory whenever either Professor Jao or Chinese culture is mentioned.

"More people should endeavour to make a study of Professor Jao's scholarship. It provides a profound window into traditional Chinese culture. Only when people understand where they are from can they know where they are going. Without their cultural roots and legacy, individuals and countries will not gain long-term development and progress."

Whenever he could find the time, Simon would reread and review Professor Jao's calligraphy, paintings, and monographs.

"Professor Jao was able to apply his knowledge and scholarship in crafting his paintings and calligraphy. He would share the history of painting, starting with an introduction to Zhao Mengfu before teaching me how to appreciate various landscape paintings. His lectures were fascinating and memorable. Now when I look at Professor Jao's landscape paintings, I can tell that they are masterpieces. They are uniquely powerful, refreshing, fantastic, and classical," Simon said, pointing to one of Professor Jao's works in the collection. "The composition is simple, containing a few willow trees and vast landscapes, but the artistic conception is profound. The few strokes are well-spaced. The painting is rich in meaning. The lines showcase his solid foundation in drawing. The implication is sincerely conveyed. Great and unique landscape paintings have a broad meaning, fully reflecting a scholar's character and noble sentiment."

"Simply appreciating Professor Jao's paintings has already made me feel how remarkable he was, but now as I also pick up a brush to draw, I realize even more his great achievement," Simon said.

"His artistic creations are not only based on technique and skill but also his knowledge. Thus, his paintings display various topics and unlimited changes. I have been with Professor Jao for many years, and often heard him talk about landscape paintings, which are wonders of the past and present, with righteousness filling the universe. It is not until these days that I dare say I can understand a little of those paintings. It is the same with calligraphy. I used to accompany Professor Jao to Chaozhou and had the opportunity to see him write in zhuanshu, a style of Chinese calligraphy often used on seals. When I look back, I can tell Professor Jao was indeed a master of calligraphy, which was very much related to his in-depth study of philology.

Simon often thinks of Professor Jao. For this year's Chong Yang Festival, he wrote a prose poem, *Reminiscences on Chong Yang Festival*, in memory of Professor Jao, and then copied the line in xiaokai, a regular script used in Chinese calligraphy.

時不我與，惜時如金，生命短暫英名永存。疫情轉虐，書畫怡性。遠望雲空清澈，參悟多彩生活，心坦然對風雲際會。茶香綿長，斗室懷恩。凝神家中屏風，黃花梨上饒畫，梅蘭竹菊總關情。今日重陽，永念饒公。東方鴻儒，德高望重，博學睿智，兼容並蓄，博覽群書，執著學問，淡泊名利，一生踐行"求真、求是、求正"。智者之風，山高水長，斯人已逝，其品德、修養、睿智、淵博，至今垂範後世。激勵來者，永以為志。

❖ *Simon's manuscript, "Reminiscences on Chong Yang Festival"*

《重陽感懷隨筆》

今日重陽節，於家中飲茶。

靜佇於窗前遠眺，陽光格外嬌艷。秋景美極

此爲絢麗成熟之季節，令陶陶然。

沉浸於一己思緒，思忖漫步黃昏，

獨對絢麗夕陽紅。

正如毛公詩：「人生易老天難老，歲歲重陽，

今又重陽」，物是人非，天無情，天不易老；

*Today, I was having tea at home on Chong Yang Festival, also called the double ninth festival.*

*I was standing quietly in front of the window, looking out and far, and saw the sun was shining brightly. Such autumn scenery is beautiful! This is a colourful season of harvests, which makes me happy.*

*I was immersed in my thoughts, thinking about walking in the dusk, alone with the gorgeous sunset.*

*As Mao Zedong's poem says, "Humans age too easily, but not nature; The Double Ninth comes every year, just as today." Nature stays unchanged, but humans are no longer the same.*

*The lost time will never be found again, so humans must cherish their remaining time like gold. Life is short, but reputation will be everlasting. When the pandemic is raging, painting and calligraphy can nourish the soul.*

*Looking at the blue sky with some clouds, I reflected upon my colourful life and felt calm and relieved in the face of crisis. The fragrance of tea lingers, the smell of burning incense stays long, and the room is full of grace.*

*I gazed at the folding screen at home, which was made of Chinese rosewood. Above the screen was Professor Jao's painting, the plum, the orchid, the bamboo, and the chrysanthemom in the scene carried forward a great emotion. On the day of Chong Yang, I will never forget Jao Gong. He was a great scholar in the East, with high moral standing and prestige.*

*He was learned and wise; He was inclusive, having read many books and was dedicated to his studies. He was indifferent to fame and fortune. He had been practicing "truth-seeking, fact-pursuing and value-persisting" throughout his life.*

*The characters of a wise man will last forever. Even though Professor Jao has passed away, his virtue, character, wisdom, and profoundness have served as a model for succeeding generations and will continue to inspire those future scholars. He will always be remembered.*

"After all these years when I spent with Professor Jao, his knowledge and character, personality and status have shown me what it means to be a great scholar and a great man. He has significantly impacted me. Professor Jao taught me the value of traditional Chinese culture; it was also Professor Jao who enlightend me to look at the world with a generous attitude, an open mind and to be in the spirit of giving to others. That's why my idea has always been very simple: do what I can to promote traditional Chinese culture."

A year after the establishment of the Jao Tsung-I Petite École, Simon and two other gentlemen made an effort to found the Jao Tsung-I Petite École Fan Club. Simon became the Founding President and has since helped raise fund and find resources to develop the Petite École. The Fan Club has supported around 300 events in the past decade, including seminars, lectures, exhibitions, and publications. In 2013, the first Academy of Sinology named after Professor Jao Tsung-I was established in Hong Kong, and, in the spirit of Buddhist giving, Professor Jao once again donated a number of paintings and calligraphy to raise funds for the development of the Academy. Simon was the Founding President of the Development Committee and has contributed funds and invested much effort. In his speech at the ceremony, he began by expressing his respect for the Professor's kindness and thanking him for donating his collection of works to academic institutions many times over the years to support the development of traditional culture studies.

"This is what Professor Jao inspired me most, giving is the most precious act. If people are not willing to give to society, they will eventually lose. What is gained from society can also be given back to society. That is our responsibility. Many people only talk the talk, but they don't walk the talk. In his life and research, Professor Jao had always upheld the principle of 'truth-seeking, fact-pursuing and value-persisting', and he never deviated from that spirit. Such philosophy affected my life comprehensively."

On Simon's desk is the inaugural issue of Bulletin of the Jao Tsung-I Academy of Sinology. Professor Jao's drawing of a lotus in full bloom is on the cover.

慈悲
喜

❖ *In 2004, Simon spoke at a donation ceremony held by the Cultural Affairs office of the Canadian Embassy in China.*

CHAPTER

# X

# RECIPROCITY

/

## CARING FOR SOCIETY

# From Society, For Society

Victoria, the capital of British Columbia, is the oldest city on the Pacific coast of Canada. Walking east along Fort Street downtown and following the Gold Coast Art Centre signs for a few minutes, you will arrive at the Great Victoria Art Gallery, hidden by lush foliage. The mansion originally built in 1889, has since been renovated, but the Victorian style architecture remains in intact. The building is asymmetrical, the layout of the courtyard carefully arranged, the pavilions, corridors, and walls classically and delicately decorated and carved. In 1951, the owner donated the building, which was promptly retitled the Great Victoria Art Gallery, the museum officially opening to the public that same year. Nearly half a century later, the Gallery became the first stop on the tour of the Ancient Chinese Jade Art Exhibition in Canada.

On September 28, 2000, the Great Victoria Art Gallery was filled with visitors. Even the unexpected autumn chill did not dampen the enthusiasm.

After the successful showing in Victoria, the exhibition travelled to the Museum of Vancouver, the Art Gallery of Hamilton, Winnipeg Art Gallery, Edmonton Art Gallery, the Museum of Civilization of Quebec, the last stop being the National Gallery of Canada. The tour lasted two years and seven months until mid 2003, spanning seven cities in five Canadian provinces. All the exhibitions received unanimous praise from the local communities.

"The rich Chinese culture of different dynasties has allowed us to see the outstanding achievements of humans in art over the last thousands of years," Jean Chrétien, the then Canadian Prime Minister said. "The exhibition features jade carvings from the Neolithic period to the Qing dynasty, and represents the first time the treasures have been exhibited outside China. The craftsmanship and beauty of these jade works are breath-taking. It is a rare opportunity for visitors to witness these exceptional works."

Adrienne Louise Clarkson, the then Governor General of Canada, spoke, "I believe that everyone who visits the jade exhibition will leave with a sense of enlightenment. China has contributed infinitely to the world's art treasures."

Fan Shimin, the Director of the Art Exhibitions Centre for the Circulation of Cultural Relics, one of the co-organizers of the tour, was delighted with the tremendous success of the exhibition. Fan had worked at the National Museum of China and once served as the deputy director of the International Friendship Museum. Fan had spent the vast majority of his professional life in museums as well as other culturally related fields and was about to retire and looked forward to the event as a crowning achievement in his career, yet he never imagined that the days after the exhibition tour would be "the most difficult period" in his life.

Fan is quiet and gentle, speaking at a pace and with elegance appropriate to a well-educated man. "Those top-level exhibits, such as the gold-threaded garments and some important artifacts, were borrowed from museums in Hebei and Guangdong. It was specified that a fee had to be paid for borrowing the objects."

"The signed contract specified that the Canadian Conservation Foundation would bear the costs. However, Dr. Nelly Ng, the President of the Foundation at the time, could not raise enough funds to pay the costs after the exhibition tour, as economic conditions in Canada were not altogether positive. The staffs at my centre were very worried because the money could not be paid without receiving funds from the Foundation. We were very anxious."

Dr. Ng was of medium height, with short hair and an appealing smile. Wearing a pair of gold-rimmed round glasses, she appeared intelligent. She was called Dr. Ng at the Foundation because of her career as a general practitioner.

Talking about the exhibition, she recalled, "We started to plan this exhibition in November 1998; then, the Foundation signed the China-Canada Agreement of Cultural Heritage in the Great Hall of the People. After that, the plans for the exhibition were carried out. The exhibits were carefully selected by the Art Exhibitions Centre for the Circulation of Cultural Relics of the National Cultural Heritage Administration, the co-organizer of the exhibition. Jade pieces from prehistoric to modern times were collected from all over China, making the exhibition an eventual possibility."

"The contract stipulates that the museums in China would be repaid after the successful completion of the tour. It was very embarrassing not being able to deliver what was promised," Fan Shimin still remembered.

When the payment was delayed, those institutions began to call the centre's superior administrative unit, the National Cultural Heritage Administration, to inquire about the matter. There was some speculation that perhaps, the delay was due to some malfeasance on Fan Shimin's part.

"There was some skepticism about whether it was true that the payment couldn't be made in time? Or had I done something else with the money?" Fan Shimin said. "The exhibition was a success, but for a long time afterwards, I was under tremendous pressure," Fan Shimin said in a lowered voice.

"This matter must not be left unresolved," Fan Shimin thought. "No matter what, I have to settle it before I retire."

"So, I told Dr. Ng that I hoped she would do her best to get the funding and send it to us."

In November 2003, he received a long-distance call from Dr. Ng, who sounded relaxed and happy.

"Mr. Fan, there is no need to worry anymore. I found a Hong Kong businessman. He is a very good person and will sponsor our exhibition. His name is Suen Man, same as the Founding Father of the Republic of China." In earlier years, people used to call Simon Suen Man.

Putting down the phone, Fan was greatly relieved. The stone that had lodged in his heart for more than half a year finally fell to the ground. A few months later, he met Simon for the first time at the donation ceremony held by the Cultural Affairs Office of the Canadian Embassy in China.

"I still remember it was an early spring evening when Simon stood on the podium and gave a speech in English. He was thin but had a strong voice. He spoke about his donation to support cultural exchanges between China and Canada, specifically the Canadian tour of the Ancient Chinese Jade Art Exhibition. The contribution would be forwarded to the Art Exhibitions Centre for the Circulation of Cultural Relics through the Canadian Embassy in China.

"You can't imagine how happy I was!"

Even after all these years, Fan Shimin still grows excited when talking about that night.

"I thought I was so lucky to meet this kind-hearted man. Although we barely knew each other before, he donated a huge amount of money to help me out of the dilemma. All the suspicion surrounding my role had disappeared. I can now retire without any worries." Fan described his first impressions of Simon sincerely, "In my eyes, he is a living bodhisattva who is generous and passionate about culture."

After the ceremony, he came forward to express his gratitude to Simon at the buffet reception. Simon was not aware of all the anguish Fan had experienced or the many memories that Fan had not only of the contribution but about the arduous steps leading up to the momentous ceremony. The two just exchanged a few words out of courtesy. That was it.

Likewise, Dr. Ng attended the donation ceremony. In her opening remarks, she said straightforwardly: "The exhibition was a great success, and it was significant. It introduced China's rich and long-standing culture and enriched the Canadian's knowledge of China. There is one person behind this success that we all should remember. Without his support and help, the event would not be completed, and there might even have been trouble."

By "trouble", Dr. Ng was referring to the settlement of payment. And that benefactor was of course Simon.

"It is very worthwhile to promote Chinese culture and art. Nevertheless, it is always hard to seek financial support because only a few are willing to help. As an entrepreneur, Simon understands the power of culture, so he is willing to give. This is a rarity."

When she first met Simon in Toronto, Dr. Ng had held an exhibition of Chinese paintings by famous artists of the 20th century in three cities across Canada. Here, she learned of his interest and love for traditional art and culture. A few years later, when she met with Simon again in the same city, she told him about the funding issues regarding the jade exhibition and asked him if he could help. "Simon said yes without any hesitation," she remembered.

"He also supported the Canadian Can4Culture Foundation to exhibit the master artist Wu Guanzhong's works at the Canadian Museum of History."

Located on the banks of the Ottawa River, the Canadian Museum of History, founded in 1856, is one of the most popular museums in Canada. Dr. Ng mentioned that one year Simon passed Ontario on his way to the museum, Gerry Phillips, a cabinet minister, took the time to meet with Simon in order to thank him for his contribution to the cultural exchange between Canada and China.

Simon became acquainted with Master Wu Guanzhong through this exhibition.

"I went to Beijing with Dr. Ng and visited Master Wu at his home. At that time, Master Wu had refused to attend public events, not to mention receiving guests at home. We felt much honoured," Simon recalled.

The disparity between Master Wu's outstanding artistic achievements and the simplicity of his life was striking. This contrast impressed Simon greatly.

He remembered: "His house still had a cement floor and looked in need of renovation."

Master Wu remarked that his house was perhaps alone in all of Beijing in such a state. Simon recalled that Master Wu was a very modest person.

About a year after retirement, Fan was appointed as the Vice President of the China Museums Association. After assuming his duties, his first task was to revise *The Annals of China Museum* in order to support China's bid to host the 22nd General Conference of the International Council of Museums in 2010.

"It has been ten years since *The Annals* was published in 1995 nationwide. During the decade, the number of museums has more than doubled, and the collected items have been more diverse. For this reason, Fan noted that *The Annals* should be revised.

"However, funding is needed for the initiation, preparation, and implementation of the revision. The China Museums Association is a non-profit organization and is not able to support the endeavour. Where can we get the money?" Fan, as the executive editor in charge of forming the editorial board, became very worried.

He suddenly remembered the words of Dr. Ng when she introduced Simon two years ago, "Simon is very supportive of cultural causes."

Fan decided to contact Simon who enthusiastically invited him to a meeting in person.

"That's when I really got to know Simon: by visiting his business in Dongguan, talking to him face to face, having dinner, and discussing with him our situation," Fan said.

"He was very supportive of *The Annals* and immediately decided to allocate 1 million RMB as start-up funding; since then, he has made several more contributions to support the compilation."

Dr. Ng described Simon as "straightforward", "generous" and "passionate about culture." Fan couldn't agree more.

Built in 1420, the Shufang Zhai, the Lodge of Fresh Fragrance, is located east of the Chonghua Palace in the Palace Museum. The exterior and interior of the building are well preserved, but the place is not open to the public, but only to receive VIPs in the Palace Museum or hold significant events.

Green pillars and a vermilion fence surround the theatre on a raised platform one metre above the ground. A plaque is affixed between the double eaves on the front of the structure. The words enamelled in green and gold read "Sheng Ping Ye Qing", meaning peace and prosperity. The couplet on the left and right of the plaque reads, "The sun shines on the Yao Terrace, and the world is at rest and the drums blow; the wind is clear, the time flows, and the joy of the world is written in song." During the Qianlong period, the emperor would visit Shufang Zhai every year at the time of the Lunar New Year to inscribe words of blessings.

On October 12, 2006, the initiation ceremony of the 2006 edition of *The Annals of China Museum* was held at Shufang Zhai.

The event was co-organized by the China Museums Association and the National Palace Museum. At the ceremony, Simon was appointed as the Honorary Editor-in-Chief of the 2006 edition of *The Annals of China Museum* and the Honorary Vice President of the China Museum Association. The guests attending the event included Zheng Xinmiao, the then Vice Minister of Culture and Director of the Palace Museum; Zhang Bai, the Deputy Director of the National Cultural Heritage Administration; Zhang Wenbin, the Chairman of the International Council of Museums China; Li Wenru, the Deputy Director of the Palace Museum; Ma Zishu, the Executive Vice President of the Chinese Society of Museums together with the directors of several museums, representatives of the cultural and museum

❖ *Photo taken on October 12, 2006, at the ceremony at the Chonghua Palace and Shufang Zhai of the Palace Museum. From left: Li Wenru, Wang Hongyi, Ma Zishu, Zhang Bai, Simon, Zhang Wenbin, Zheng Xinmiao, Shu Yi, Chu Hok Ting, and Li Fusheng*

community, and friends from various cultural circles. China's Central Television (CCTV) covered the event during the evening news program.

Simon delivered a speech at the ceremony: "The Chinese people have a 5,000-year history of civilization. Therefore, our ancestors have left behind many precious historical relics for future generations. These antiquities are collected in museums. They are the soul of the nation and the witness to the glorious history of the Chinese people. The compilation and revision of *The Annals of China Museum* contribute to the preservation of the common cultural heritage of humankind. As an entrepreneur, I am willing to make my modest contribution to the development and prosperity of the cultural and museum causes."

Simon once mentioned to Fan his youth in which he had not studied much during that period and was not highly educated. He hoped that young people would not again experience the lack of education in traditional culture: "We should value culture, its origins and continuous development. Any individual, or any nation, must not ignore the power of culture."

In the Palace contained six hundred years of history and recorded the splendid culture of the Chinese nation, Simon expressed his respect for the power of culture directly. Dressed in a black tweed suit, he spoke simply but sincerely.

Four years later, the first batch of *The Annals of China Museum* went to press. Simon wrote a congratulatory message for the publication: "Although I am in Hong Kong, I have always been concerned about the compilation of *The Annals*. When I learned that the first batch of *The Annals*, including the Zhejiang, Guangdong, Hong Kong, and Macao volumes would be published soon, my heart was filled with joy. Here I send you my warm congratulations. Though having not participated in each step of the compilation of *The Annals*, I have been following your work progress all the time."

"That was true. Whenever he was on a business trip to Beijing, and no matter how busy he was, Simon would always make time to visit the editorial office," Fan said, taking out a photo of Simon sitting at a long table. He was wearing a short-sleeved T-shirt with a beige lapel. A navy-blue sweater was taken off and placed over his shoulder. There was a significant difference in temperature between the north and the south, so traveling from HongKong to Beijing is like traveling across the seasons. The background of the photo was a painting of undulating mountains.

"The editorial team took the picture when he visited the office. Simon liked it very much, and it was later used as an illustration for his congratulatory message in the publication." In his congratulatory message, Simon also referred to the photo, saying that "the majestic mountains evoke images in people's minds and give them strength and power." From that observation, Simon remarked on "the dedication and seriousness of the editorial team in compiling the work and believed that they would not back down no matter how many difficulties they encountered."

At the same time, the bid to host the $22^{nd}$ General Conference of the International Council of Museums in 2010 was successful. In the summer of 2007, Simon, turning 50 in a few months, was invited as a member of the China Museums Association delegation to Vienna, Austria, to participate in the flag presenting ceremony for the Conference.

At the ceremony, the delegation members took a group photo under the flag. Standing in the middle, Simon was full of high spirit and joy. Having worked in the commercial world for many years, he has long ago become skilled at maintaining a placid expression on important occasions, but at this moment, he glowed with happiness.

*In 2007, Simon was a member of China Museums Association delegation to Vienna, Austria, to participate in the flag-presenting ceremony.*

In November 2010, the General Conference was held, as scheduled, in Shanghai. As Honorary Vice President of China Museums Association, Simon attended and witnessed this critical event in the history of museums. At the suggestion of the China Museums Association, the theme of this international event was "Museums for Social Harmony." A few years later, he founded Simon Suen Foundation and later set up Sun Museum, advocating the same philosophy of "culture for harmony", believing that culture can have a positive and significant impact on the development of society.

"He is an unusually sincere entrepreneur who respects traditional culture; he understands the significance of *The Annals* and agrees from his heart that people from all sectors of society should care for and support its compilation."

Having grown to know Simon well, Fan speculates that his global business has shaped Simon's views on culture. Fan remarked in this vein that "possibly because he is running a multinational company, I think Simon holds an international vision in matters of culture."

❖ *Simon received a book collection about the Palace Museum when he visited the then Director of the Place Museum, Professor Zheng Xinmiao.*

"Before the publication of the second edition of *The Annals*, we invited him again to write a congratulatory message. Simon wrote: '*The Annals* are a medium, used to record the development of museums in China objectively and to sort out and preserve the Chinese culture for our own. Meanwhile, *The Annals* can also show the world the long history of the Chinese nation. People worldwide can understand China historically and comprehensively. Thus, we can strengthen the communication between China and the world, enhance our many friendships, develop a spirit of cooperation, and eventually achieve harmony."

"He has a very profound knowledge and understanding of culture," Fan said emotionally, after reading through the preface composed by Simon for *The Annals*. At the end, he added, "This profound understanding may have grown out of his friendship with Professor Jao Tsung-I."

When Simon and Professor Jao Tsung-I had known each other for ten years, Simon arranged a special dinner at the Peninsula Hotel in Hong Kong. The theme was "Getting to Know Professor Jao, the Master of Chinese Studies." Fan was invited to attend the dinner and share on stage as a guest speaker: "Once I went to the Palace Museum with Simon to visit the director Professor Zheng Xinmiao. I clearly remember what happened. Simon asked us in Mandarin with a Hong Kong accent, 'Do you know Professor Yao?'"

In Cantonese, "Jao" sounds similar in pronunciation to 'Yao' in Mandarin; Simon, with a Hong Kong accent, mispronounced the word as 'Yao'. Director Zheng did not understand and asked me, 'You knew Simon better. Do you know who this Professor Yao is?' I shook my head and said I didn't know either.' It was only when Simon pronounced 'Rao', the right pronunciation of 'Jao' in Mandarin, that the Director and I suddenly grasped Simon's meaning." Fan smiled, "But Simon's Putonghua has gotten better since."

The audience burst out in laughter.

Professor Jao and his family attended the dinner party, during which his younger daughter, Ms. Angeline Yiu, gave a speech.

Wearing a navy-blue suit jacket and a light pink and purple plaid scarf, Ms. Yiu is a petite woman who looks quite a bit like Professor Jao. Although she speaks softly, her voice was gentle and powerful.

She said that in 2002, when Jao Tsung-I Petite École was being prepared, a lot of things needed help and urgent support. It was by chance that she made acquaintance with Simon. Ms. Yiu summed up with a phrase Simon's support towards the development of the Petite École, Jao Studies, Sinology, Chinese traditional cultures and arts in the decade that past: "It has never stopped."

She continued after a short pause and thanked Simon profusely for his efforts on behalf of Professor Jao and her family. But that event did not end Simon's support for Chinese studies.

It was the first Friday of September 2019 when the celebration party for the 15th Anniversary of the Founding of the Jao Tsung-I Petite École Fan Club was held at Rosewood Hotel, a brand new six-star hotel by the seashore of Tsim Sha Tsui, Kowloon. At around 2 PM, just past lunch time, the staff gradually appeared outside the Grand Ballroom and displayed carefully Professor Jao Tsung-I's eight sets of calligraphy and painting works for the fund-raising auction section that night. Simon donated two of the larger works. He had treasured these works since making a winning bid at an auction house many years ago.

Though the preparation team had arrived early, it was hard to say if guests would be able to make the event on time. Luckily, when Victoria Bay glistened with lights, the guests from all fields, many key figures in cultural, political, and business worlds, were able to make the proceedings. After signing up at the reception desk, the honoured guests strolled down a black-and-white marble stone-patterned interior conceived by Tony Chi, a world-class Chinese designer before passing through an arch where the guests were transported from the Art Dec style hotel to an elegant and poetic space, a home for Chinese art. Most works displayed outside the Grand Ballroom were framed with golden wheat ear patterns and red backing paper, the pair symbolizing happiness and harmony in Chinese culture.

As the dinner party started, a video clip cherishing the memory of Professor Jao was played. With the slow and soothing sound of guqin, it felt as if guests had been brought to a peaceful garden under the moonlight. Simon became lost in thought. It was a totally different scene during the auction section. The atmosphere was warm, and many works were sold above the upset price. Under the uniquely shaped crystal chandelier, the sound of applause lasted for a prolonged period when at the close of the night, it was announced that the eight-digit-amount had been raised for the charity. Professor Jao's *To a Bright Future* was displayed on stage as a virtual backdrop. Simon had donated the original painting.

The painting was divided into upper and lower spheres. The upper portion, finished in the style of Dunhuang grotto murals, showed a pair of golden birds, which were flying. Six flourishing pines drawn in the Professor's inimitable style occupied the lower half of the painting. The splendid work commanded the highest bid of the night. Together with another painting he had donated, Simon had contributed nearly 10 million HKD to the proceeding. In her closing remarks, Ms. Angeline Yiu, first thanked Simon, the President of the Organizing Committee, at the very first.

According to the folklore in China, the number "9" stands for chaos. While numerology may be an unreliable prognosticator, in this particular case, international economics of 2019 were indeed tumultuous. The intensified trade war between China and the United States had impacted all industries. The tensions in Hong Kong since June, merely added to the unstable mix.

In such a harsh climate, many activities had to reschedule or be cancelled in their entirety. In this context, it was very challenging to choreograph a party that was befitting such a master as Professor Jao. Although the dinner party was only three-hour in length, Simon and the whole Organizing Committee had laboured for 13 months to plan the event. Ms. Yiu understood the many behind the scenes difficulties and the time and effort that Simon had devoted to make the event come true.

In mid 2018, after Professor Jao passed away, the Fan Club held its first annual board meeting. After reading over the financial statement which showed a deficit, Ms. Yiu grew anxious. A long silence descended over the meeting room. The steaming hot tea started to cool. Simon signaled the staff to have the tea replaced and said, "Though Professor Jao passed away, the work at the Petite École must go on, and we must solve the financial problem." His tone was determined, caring, responsible, and as supportive as it was in 2003, when organizing the Fan Club, supporting the development of the Petite École.

"It's the $15^{th}$ anniversary of the founding of the Fan Club in 2019. By gathering for that occasion, we can cherish our memories of Professor Jao, and to spread his academic achievements to a wider audience and to allow others to appreciate his spirit."

Simon has a profound understanding of the significance of this endeavour. A great deal of tedious work went into the planning and arrangements for the event. Once the date of the dinner party was set, Simon spared no effort to invite friends from all fields, encouraging their participation and support. Meanwhile, Simon held regular meetings with the team, following up with the preparation work in a nuanced manner from sourcing the items for auction, designing the menu to inviting master of ceremonies.

Ms. Yiu said, "Simon plans the event full-heartedly and affectionately."

Enjoying the delighted and peaceful atmosphere at the dinner party, the guests were whispering and chatting with each other. The latecomers were led, under guidance, to their seats. The Grand Ballroom was filled with activity until, at last, Simon walked onto the stage and said energetically, "Good evening, everyone."

The room became instantly quiet.

He explained in his speech that it was very hard for everyone to gather for the party, and he was more than glad and honoured to be able to make it happen with fellows from the Fan Club and the Organizing Committee. He described himself as a dream chaser led by Professor Jao, pursuing the dream of the revitalization of Chinese culture, and hope more people to join together in making this dream a reality.

Simon, no stranger to oration, is skilful at giving concise, rational, and reasonable answers to questions and in general, making impromptu speeches. He elaborates his main point well by sharing his personal experiences and feelings in a relatable manner. The dream chaser standing at the stage had the same determined look in his eyes as 15 years ago, when he served as the Founding President of the Fan Club, although his temple hair was turning grey.

It is relatively easy to stick with an action for 15 days or 15 months, but it is an extraordinary challenge to persist for 15 years. Persistence, the key to answering many questions, is impossible without unwavering faith and righteous thoughts.

Simon was the last to leave when the dinner party ended. The lights glimmering from Victoria Bay surrounded him. The 280-metre-tall building decorated with limestone and having brass exterior shed a golden halo, circulating behind him. The lights and halo accidentally fitted the theme of the night — *Glow with Jao Studies*.

Since meeting Professor Jao in 2002, Simon had led and organized many events, devoting time, efforts, and financial supports to making the Professor's dream come true. He said frankly that it was well worthy.

"I have been privileged to learn a great deal under Professor Jao's guidance. As many foreigners think highly of their history and culture. We, Chinese, should not ignore and forget our own traditional culture and art."

In the summer of 2010, the Fan Club funded an event called The Fragrance of Mogao — Art Exhibition of Jao Tsung-I's Dunhuang Painting and Calligraphy Works. Simon and seven presidents of the Fan Club went to Gansu Province, with Professor Jao for the opening ceremony. The co-organizer of the event, Dunhuang Academy, gave a special introduction of the desert culture for many guests who were

❖ *Simon gave a speech as the Founding President on the inauguration ceremony of the Fan Club in 2004.*

❖ *Simon gave a speech as the President of the Organizing Committee on the 15th anniversary of the founding of the Fan Club in 2019.*

first time visitors to the province.

Wearing a sun hat and sand-proof shoe covers and riding a camel under the sun for the first time was indeed a special experience, but what touched Simon most was the magical Mogao grottoes, which to him, represented a desert treasure.

Due to the harsh natural environment, the destruction from pollutants and other man-made elements, and a lack of maintenance, it was estimated that the Dunhuang grottoes will vanish in another 300 years. After listening to the introduction about the conservation work carried out by the Dunhuang Academy, Simon and the other seven presidents of the Fan Club decided, on the spot, to donate 200 thousand HKD respectively to the China Dunhuang Grottoes Protection and Research Fund for the preservation and conservation of the grottoes.

After returning to Hong Kong, Simon worked closely with colleagues from the Fan Club for two months and held the Focusing Dunhuang — Dunhuang Themed Fund-Raising Dinner that October. The event raised over 13 million HKD. The Director of the Dunhuang Academy at the time, Professor Fan Jinshi attended the event and thanked the donors for their "loving hearts", which would greatly encourage the work in support of the restoration and digitalization of Dunhuang

❖ *Simon was awarded with a certificate of gratitude for organizing the Focusing Dunhuang — Dunhuang Themed Fund-Raising Dinner. Left: Professor Lee Chack Fan*

grotto murals. Simon went to Gansu later on a duty travel in 2016 and joined in a reunion with Director Fan. The two of them chatted and took photos together as old friends. And it was only then that the local officials who joined the trip discovered Simon's support for the heritage conservation in Gansu.

In addition to supporting the development of the Fan Club and Petite École, Simon is working to promote Jao Studies on an international scale by establishing Jao Studies Foundation to encourage institutions around the world to study Professor Jao's achievements in academic, literature, art. The aim was, to promote the breadth and depth of the Professor's thought on a global level. In 2012, Simon in his role as the Organizirg Committee President of the celebration ceremony for the establishment of Jao Studies Foundation raised 75 million HKD.

In 2013, a year later, the Jao Tsung-I Academy of Sindogy, the first institution of its kind named after Professor Jao was founded at Hong Kong Baptist University. Professor Jao, nearly a hundred years old by that time, was able to attend the ceremony. Throughout the event, Simon was always around the Professor, who held his hand tightly. Wearing his red tie, Simon could not be happier that he had made it a reality. Simon's care for education and traditional culture became combined in the activity. He served as the Founding President of the Development Committee of the Academy and the Co-President of the fund-raising dinner party in October later that year. Through Simon's and committee members' efforts, the dinner party was a huge success, raising over 70 million development funds for the Academy. In his speech, Simon expressed the profound expectation to the Academy, to carry forward the conception of "Eastern Learning Spreading to the West" of Professor Jao, to become a bridge between Chinese studies and the world, as well as to promote the understanding of Sinology in the world.

Simon's motive in promoting Chinese traditional culture are sometimes misunderstood and is, therefore, bearing noticeable pressure behind the curtain. However, he has calmly handled the idle speculation, maintaining throughout his typical positive attitude, "My initial goal, to encourage more to study traditional culture and to spread the culture, has never wavered."

In his many interviews, Simon has reiterated his profound commitment to conserving traditional culture. He believes it is his obligation to give back to society.

Not long after the Academy was founded, Simon, as well as representatives from Hong Kong culture and education circles, visited the Institute of Chinese Studies at Peking University. This was the first time that Simon had visited the university. In

❖ *Simon at the Inauguration ceremony of the Jao Tsung-I Academy of Sinology at Hong Kong Baptist University in April 2013. From left: Mr. Wang Guohua, Professor Yin Xiaojing, Professor Xu Jialu, Simon, Professor Jao Tsung-I, Professor Albert Chan, Mr. Eddie Ng, Professor Yuan Xingpei*

❖ *Simon delivered a speech at Jao Tsung-I Academy of Sinology fund-raising dinner.*

the classroom, eight desks were put together to form a conference table, with hosts and guests sitting face to face. Professor Yuan Xingpei, the Director of the Institute, hosted the meeting. Executive Deputy Director, Professor Wu Tongrui and the other professors of the Institute including Professor Yan Jiaming of Archaeology, Professor Zhang Chuanxi of History, Professor Sun Qinshan of Archeography, as well as the postgraduate students of the institute, attended the meeting.

Professor Wu first introduced the achievements of the Institute in research and teaching over its 22-year history. While strolling along the bank of the Weiming Lake with Professor Yuan before the meeting, Simon recalled a student from Hong Kong that the Institute had admitted, Louis Cha. According to a common story, Louis had once struggled with a difficult question on history and finally got the answer after consulting with a professor from the Institute, then decided to apply the Institute, becoming Professor Yuan's student.

"Cha is a man with taste. This is a great place to do Chinese studies, with such a great view and so many experts." Everyone laughed. Simon continued, "Many universities are not valuing Chinese studies enough because of students' future employment. Graduates of programs like finance, economy, and IT are welcome in the job markets, while it is much harder for graduates of Chinese Studies and humanities programs. Even if they can secure a job, their salary will be lower." Professor Yuan, who sat opposite Simon, smiled and added, "Much lower."

"That's not right." Simon said in his much-improved Putonghua, "I am an entrepreneur myself. Although the company is not gigantic in size, culture is very important as an international company. Employees from different regions have different backgrounds. Understanding cultures besides their own is vital. Only in this way, can they get along with each other. Unlike the other three ancient civilizations which became ineffectual, we stand resolute because of our firm cultural foundation. We should not look down on Chinese studies program just because facing difficulties in job market or the salary level."

Simon paused and said he was delighted to see so many students working on Chinese Studies here. Having been enlightened by Professor Jao for many years, he understood the importance of Chinese studies. For that, he admired professors' and students' concentration and their shared commitment to promoting the subject. By the end of the meeting, when taking the group photo, he insisted that the professors sat in the middle, while he was on the side, smiling joyfully.

During the dinner afterwards, Professor Yuan talked about a student of his, who

❖ *Simon visited the Institute of Chinese Studies at Peking University in 2014. Front row from left: Professor Chen Zhi, Simon, Professor Albert Chan, Professor Yuan Xingpei, Professor Sun Qinshan, Professor Yan Wenming, Professor Zhang Chuanxi, Professor Wu Tongrui, and Dr. Thomas Tang.*

had come from the remote area of Guizhou Province. He hoped that the diligent and talented student could study abroad and return to the University in the future. As a teacher who loves and cherishes talents, Professor Yuan became emotional and wet his eyes. Hearing this story, Simon replied instantly that he could, if need be, provide financial aid. After returning to Hong Kong, he urged his daughter who oversaw the foundation to follow up and implement his words.

The student was not the only one who received financial assistance from him. Simon had lost count of how many students from poor families that he had provided financial support towards their pursuit of education. Since 2009, he started to make donation to the Virya Foundation and funded financially challenged college students from six universities in Mainland China. In 2013, Simon funded the Young Artist Development Foundation to help talented teenagers from poor families to undergo professional art trainings. Afterwards, Simon set up his own foundation and established different scholarship programs, encouraging students to dive into the study of traditional culture.

When Simon was asked why he pays special attention to education during an interview, he said, "In additional to encouraging students to study Chinese traditional culture, I sincerely hope to help poor students to have the same opportunity to be educated. Only through education and hard work can they leave poverty" This represents Simon's cultural caring on education and his focus on poverty relief.

On September 10, 2014, the *People's Daily Overseas Edition* posted an article titled *Why Some Rich Hong Kong Citizens Favour Donating Money for Education.* Here it was reported that Simon had donated 30 million HKD and established the Mr. Simon Suen and Mrs. Mary Suen Sino-Humanitas Institute at HKBU. Simon's response was typically humourous and good-natured, "It's primarily because I have chemistry with the President." Simon explained that he was definitely not a rich person and just did his outmost.

Joining HKBU as the President in 2010, Professor Albert Chan is a world known organic chemist who had studied abroad in Japan and America. He is also known for his capabilities in Chinese poetry and calligraphy, being praised by media as a poet-like scientist. Professor Chan ardently loves Chinese traditional culture, which conveys a great deal of wisdom and philosophy. After assuming his role as the President of HKBU, he intended to promote the study of culture and history at the university and talked with Professor Chen Zhi, the Department Head of Chinese Language and Literature at the time.

Professor Chen Zhi soon submitted a proposal to establish a Research Institute of Chinese Studies and Sinology at HKBU. A phone call with wonderful news reached Professor Chen at the beginning of 2011. On the other end of the phone, President Chan shared with him excitedly that the proposal was endorsed. With Simon's extensive support, the Sino-Humanitas Institute had been founded, aiming at the promotion of Chinese traditional culture and the research on Chinese classic literature and Sinology.

"What impressed me the most about President Chan was not only his insight on poetry, but also his devotion to education." Simon recalled when he first joined the University Foundation. He found an internal donation list, with President Chan's name on it as a donator.

"I was deeply touched at the time. It is rare for a president to exhibit such a selfless devotion to a society."

Likewise, Simon's generosity won him the respect and support of many in his community. In fact, his consistent devotion and benefaction over the years moved

❖ *The inauguration ceremony of the Mr. Simon Suen and Mrs. Mary Suen Sino-Humanitas Institute. From left: Professor Albert Chan, Mr. Wilfred Wong, Professor Jao Tsung-I, Simon and Mary, Professor Chen Zhi*

and inspired more people to join in this meaningful work. Simon did not take friends' support for granted but always kept in mind these valued individuals. At his 60th birthday party, he gave a speech on his principle, "from the society, for the society," that on his way there have been many people from different fields that had given him enlightenment and support at different times. He, then, read out over the course of five minutes each and every person's name and thanked each individual one by one.

In 2012, Simon's devotion gained the approval of society. Honours were received in succession. He had been awarded as Honorary Fellow and Honorary Doctor of different universities and appointed as a Justice of the Peace by the Hong Kong SAR Government. The first thing Simon did after all the awards and honours, was to share his "feelings on contributing to the society" with the colleagues. He said that it was his dream to establish sound values and to contribute to the society. The way to achieve this, in his answer, was "benevolence and devotion": from the society and for the society.

"I always think that I can have my achievements because the society gave me opportunities. So, I always try my best to give back to the society, doing something for it, which I think is my responsibility. As for paying extra attention on education and culture, initially it was because of my personal experiences. I really hope the next generation could receive more education opportunities and attach, as a result, an importance to culture. Through Professor Jao's enlightenment, I have come to see the pursuit of this work as my mission and idea."

In 2013, Simon was appointed as a member of the National Committee of The Chinese People's Political Consultative Conference (CPPCC). When Simon participated in the conference in Beijing for the first time, he accepted an interview. With red walls and green willows outside the window as background, wearing white shirt, Simon was mistaken for a scholar at the university by the journalists. After the interview, the reporter found it was not totally a mistake indeed.

Many years ago, the country boy running bare footed in the field under the moonlight, had not thought that he would be a noble and mature adult, saying these words; the teenager crying silently in desperation at night had not thought that he would fight against strong torrents, handle emotional fluctuation, and create a legend.

It requires a lot of power and courage for a man to keep his sincerity, honesty, and motivation after countless suffers and to keep a positive attitude. It all depends on righteous thoughts.

The legend starts because of righteous thoughts. The legend unfolds because of righteous thoughts. The legend spreads because of righteous thoughts.

❖ *Simon was conferred Honorary University Fellowship from the University of Hong Kong.*

❖ *Simon was conferred an Honorary Degree of Doctor by Hong Kong Baptist University.*

❖ *Since 2010, Simon started to support the Community Chest, which benefited more than 140 social welfare institutions and 2.3 million citizens in need in Hong Kong.*

❖ *Simon joined the work at the Scout Association of Hong Kong in the late 90s and has been praised by the Association for multiple times over the years for his outstanding contribution.*

❖ *Fan and Dr. Ng, attended the opening ceremony of the Sun Museum in 2015.*

❖ *Simon delivered a speech at the opening ceremony of the exhibition entitled "Xu Beihong and His Times" which was held in Sun Museum in 2018.*

❖ *Simon was awarded the Bronze Bauhinia Star by the government of the Hong Kong Special Administrative Region for his years of contribution in charity, youth development, and promoting Chinese traditional culture*

❖ *Simon attended the Jao Link Inauguration and being awarded the certificate for his contribution on the development of Jao Studies over years.*

❖ *Simon attended the launching ceremony of the shooting of a documentary series "The Great Master Jao Tsung-I".*

❖ *Simon was reappointed as member of the National Committee of CPPCC in 2018 and took a photo in front of the Great Hall of the People.*

# About This Book

Simon Suen

What I remember the most is the winding and narrow lane outside my hometown, extending into the distance in the green field. It was the only way to get in and out from Shangsha Village in those days. And it was the very way I came out from.

I grew up and had hard time in the countryside. But who from my generation has not experienced these stumbles and hardship? The strength of my origin as a country boy is that I am tough and hardy. Till this day, from the big world of a small label growing on the foundation of technology innovations, these humble beginnings affect my heart as always.

My career started with an old black briefcase, and forty years had passed unwittingly. This book is a review and summary of the past.

I have different aspirations at every stage of life: When I was just released from forced labour, what I wanted the most was a bowl of rice; When I worked as a teacher in the village, I wanted a watch and a bike from a famous brand; When I moved to Hong Kong, I wanted an apartment; When I started my own business, I wanted to be the best in the industry. Though materials indeed are what push us to move forward, our constant pursuit for ideal and ambition is the genuine and long-lasting motivation. I would not be an entrepreneur for forty years without my idea to achieve greatness.

*The Great Learning* says that one must regulate oneself and manage one's family before one can govern the country. It is the same for running a business. The ancient wisdom also offers the invaluable guidance that one could learn from, which should be revealed in our daily lives with a proper and sincere attitude — this is why this biography includes not only stories about my business but also unfolds my own self-reflections.

"Steel-liked mountains I fear not, climb over from its peak I stride." I would be more than delighted if my crumbs of experience can bring readers slight inspiration.

# A LABEL, A LEGEND

**Author:** Faye Tong
**Translators:** Hong Qian, Ge Song, Mengying Jiang, Tenglong Wan, Jia Zhang, Jiahua Bu
**Editor:** Victor Jiang
**Copy Editor:** Charles David Lowe
**Designer:** Amanda Woo
**First published in January 2023**

**Published by Joint Publishing (H.K.) Co., Ltd.**
20/F., North Point Industrial Building, 499 King's Road, North Point, Hong Kong

**Printed by Elegance Printing & Book Binding Co., Ltd.**
Block A, 4/F., 6 Wing Yip Street, Kwun Tong, Kowloon, Hong Kong

**Distributed by SUP Publishing Logistics (HK) Ltd.**
16/F., 220-248 Texaco Road, Tsuen Wan, N.T., Hong Kong


Published & Printed in Hong Kong, China

**ISBN 978-962-04-5109-6**